Forest and Woodland Trees in Britain

Forest and Woodland Trees in Britain

JOHN WHITE

Westonbirt Arboretum,
Tetbury, Gloucestershire

Oxford   New York   Tokyo
OXFORD UNIVERSITY PRESS
1995

Oxford University Press, Walton Street, Oxford OX2 6DP

Oxford New York
Athens Auckland Bangkok Bombay
Calcutta Cape Town Dar es Salaam Delhi
Florence Hong Kong Istanbul Karachi
Kuala Lumpur Madras Madrid Melbourne
Mexico City Nairobi Paris Singapore
Taipei Tokyo Toronto

and associated companies in
Berlin Ibadan

Oxford is a trade mark of Oxford University Press

Published in the United States
by Oxford University Press Inc., New York

A catalogue record for this book is available from the British Library

Library of Congress Cataloging in Publication Data
(Data available)
ISBN 0 19 854883 4

Typeset by EXPO Holdings, Malaysia

Printed in Hong Kong

Preface

This book has been written for people who are interested in woods and forests rather than individual trees in the garden or in the hedgerow. Trees here are considered in a wider context that includes their history and uses. Many people who take delight in woodland would like to know more about the trees they see, or about the history of the woods and how they are managed. Those who appreciate the beauty of well-crafted wood, whether in antique or modern furniture or in a medieval timber-framed building, may also wish to know something about the nature of the wood that was used, the kind of tree from which it came, and its age when felled.

Many questions come to mind that link the wood with the tree such as, is it a hardwood or a softwood and what are the colour and texture likely to be like? To some people, the world of the living tree may seem far removed from dead wood in the timber yard. These topics are not, however, incompatible. For example, Grinling Gibbons carved in lime wood and would have used a proportion of coppice stems from ancient rootstocks as his raw material. The plants from which his poles were cut 300 years ago could still be alive, producing new wood today. Even the trees that did not survive the woodman's axe may themselves have set abundant seed. The progeny of the great oaks that have served us so well in the past live on and, with good management, produce timber of the same high quality for us today.

The aim throughout the book is to link the interests of the naturalist with those of the professional forester on the one hand, and with those of the practical woodworker and the craftsman on the other. The species described in these pages include all but the more obscure varieties to be found in British woodlands. The text and illustrations will thus enable most woodland trees to be identified, but no attempt has been made to compete with the many comprehensive guides to tree identification that are already available.

Westonbirt Arboretum   J.W.
March 1994

Contents

Introduction

Trees that are likely to be encountered in most British forests and woodlands are included in this book. These are in fact remarkably few: some 45 species and varieties make up about 90 per cent of the area of our present-day forests and woods. Of course, this does not include the many edge trees and incidental species that may form a minor part of a mixed wood, adding diversity and environmental value, but not usually having a silvicultural or wood-producing role. Selection by growers and foresters, of both native and exotic species, has been long and thorough. In some instances, their success reflects years of trial and error; in others, as with some North American exotic conifers, it is largely due to good fortune. The original batches of seed were often collected from high-quality trees growing in a natural habitat not unlike western Britain, which is where the majority of these species were planted.

The choice of tree varieties to be grown in a plantation depends on what will grow successfully in the area and on the eventual uses of the wood. In recent years survival and productivity have been increasingly considered in conjunction with the uses to which the living wood may be put, and the impact of the trees on the environment. In the past huge numbers of conifers were planted to meet the demand for softwood; the species chosen were those known to grow best on the allotted site. Different demands and different markets are now becoming apparent. For example, alternatives to imported tropical hardwoods will need to be found, some of which might be grown in Britain. Finding suitable alternatives to the trees grown at present will be difficult, although provenance selection (choosing plant material or seed from an area which is climatically and edaphically similar to the new site) and selective tree breeding have achieved considerable success. Novel species will gradually come into wider use, but this will require extensive field trials and the development of new production techniques.

The text is arranged systematically by species. For each tree or closely related group the information is arranged under five main headings, which present the same information in the same order for each species: the first section (after some etymological notes) covers the origin, history, natural range, and distribution of the species or group, with measurements of the most representative living specimens on record; the second section gives a brief botanical description; the third section is concerned with the environmental requirements of the tree and its possible impact on a site; informal principles of management follow in the fourth section; finally, in the fifth section, consideration is given to the economic value of the tree and its timber, including some fascinating historical issues which have almost been forgotten.

The illustrations follow a systematic format similar to that of the text. Most trees and forests are naturally photogenic. Any that may have a reputation for appearing somewhat dull, seen as a mass, require only a closer look to reveal some of their beauty and interest. The great storm of 1987, in so many respects a disaster, brought one unexpected benefit: it provided samples of timber from a wide range of species that would not otherwise have been available. Photographs of these naturally cleft, untreated specimens have been included here whenever possible.

Terminology used in this book (full glossary of terms on page 211)

- A *tree* is defined by Alan Mitchell (1974) as a woody plant able to exceed 6 metres (20 feet) in height and having a woody, usually single, stem.
- *Forest* has several meanings. Here the word is used in the context of 'forestry', for example, trees deliberately planted and managed to produce wood products.
- A *woodland* is an area of trees that may be managed, but is of natural or unknown origin, presumed not to have been artificially planted.
- *Timber* is, according to Oliver Rackham (1990), cut from trunks of more than 20 centimetres diameter at breast height (dbh) or 2 feet in girth at this point. (Note: in timber measurement centimetres and diameter are used as standard on standing trees and cut poles. Once sawn or otherwise processed, timber is measured in millimetres). In Britain many people prefer to measure trees in feet and inches and record girth. Breast height is taken to be 1.3 metres (4 feet 3 inches) above the ground. In the past, trees were usually measured at a height of 5 feet.
- *Wood* (or *cut wood*) consists of small poles and firewood, branchwood, scrub, and coppice. It is traditionally used for rods, faggots, small logs, and charcoal; it only includes material under 2 feet in girth (as tithes were payable on larger sizes).
- *Flowers*. The scope of this term is extended here for simplicity to also include the 'flowers' of conifers (Gymnospermae), which do not produce true flowers as such. In conifers pollen germinates directly on the ovule and not on a specialized extension of the carpel.
- *Measurements* are given in metric units, usually with approximate imperial conversions.

History of forests and woodland in Britain

Conifers were present on the earth 250 million years ago. The yew is thought to have originated at about this time, although yew fossils have so far only been dated back to the Triassic period, 70 million years later. Through the late Permian, Triassic, and Jurassic periods, conifers prevailed, at first with tree ferns or cycads (Bennetitales), and their close relatives, Maidenhair trees (*Ginkgo*), and later on with early flowering plants. Although now in decline, conifers still constitute about one-third of the world's forests. The next group of prominent trees were part of the huge class of

▲
Magnolia flower.

flowering plants (Angiosperms). The earliest examples were *Magnolias*, a genus that remains extant today. The oldest known tree fossils, which include oak, are of the Cretaceous period, which began 136 million years ago. Today's trees have remained generally unchanged for 60 million years: environmental adaptations and genetic diversification have resulted in new

◀
Old English countryside,
Herefordshire.

species, but the basic structures remain the same. The geographical distribution of trees is affected by climatic fluctuations and land movements, which have been subject to constant change. The *Nypa* palm and other tropical trees once grew in what is now Britain. In a later geological period redwoods were common. Huge deposits of much older Carboniferous plant remains—seed ferns and horsetails—are found as coal under much of the country.

In Britain the history of present-day trees and woods began about 11 000 years ago. For the past million years, the British climate has been mostly arctic with short warmer periods, each lasting some tens of thousands of years. During these spells of warmth between glaciations, plants and animals have migrated in and out of the country as sea levels have risen and fallen. After the last Ice Age, plants gradually began to recolonize Britain which had been laid almost bare by ice sheets and severe cold. Natural migration by seeding continued northwards from southern Europe until about 6000 years ago when the English Channel, deepened and widened by meltwater, cut the British Isles off from the rest of Europe. No further continental migration was possible, and the present-day native flora was determined, including only about thirty-three species of medium and large trees. Of these, some still form woodlands (including mixed woods), some are used in plantations, and others are hardly ever seen in a woodland environment.

Before 4000 BC much of Britain was tree-covered, but after that date dramatic clearances began. The phenomenon known to botanists as the Neolithic Elm Decline effectively removed trees from one-eighth of the land surface, probably through an outbreak of Dutch elm disease. Increasing agricultural land clearance continued so that by 500 BC half the land area of Britain was treeless (Rackham 1990). By AD 1200 much of the present-day rural landscape was established. Some eleventh-century examples of woods and hedges still remain. Agriculture now accounts for 77 per cent of the land surface, although urban and industrial growth, including roads, is rapidly encroaching on farmland. Due to the excessive requirements of the 1914–18 war forest cover was reduced to only 5.4 per cent in 1924 but through sustained establishment of coniferous plantations it had increased to 10 per cent by 1990. In 1984, 14 per cent of woodland was classified by D. A. Ratcliffe as 'ancient semi-natural'.

World-wide forest classification

A forest is a complex community of trees, shrubs, ground vegetation, animals, birds, invertebrates, and soil fauna. This elaborate ecosystem is dominated and protected by the trees. They reduce the impact of the wind, they intercept rain (especially deluges that might otherwise cause flooding and erosion), and they shade out direct, strong sunlight. World-wide there are seven distinct geographical zones, each having a different set of tree species matched exactly to the local climate and geography. The British Isles are near the northern limits of the 'broadleaf temperate forest zone'. The generally mild climate over most of the country allows for imported trees from several different zones to thrive in Britain. Some species will even grow better in Britain than in their natural geographical ranges. For others this advantage applies only to early growth: some imported species are capable of growing very rapidly in Britain, but a period of exceptionally cold weather will damage them or interrupt their growth. Trees accustomed to colder conditions than those normally experienced in Britain may grow well, but the wood does not always ripen properly, growth may be unpredictable, and they may suffer from insect or pathological problems. Fortunately this is not always so, and some sub-alpine species produce good, strong, home-grown timber in Britain.

▲
Serbian spruce is an endangered species in the wild, but there are small plantations in Britain.

High-latitude, mainly conifer forest (Taiga)

Geographically, Britain is slightly south of the limit of the high-latitude conifer zone which forms an almost circumpolar band of sub-arctic forest in the northern hemisphere. Taiga forest is characterized by generally low rainfall, under 200 mm, and severe winter cold (to –50 °C). In summer, by contrast, temperatures may rise to 50 °C. Pine, spruce, and larch are the major species. Many of these grow in northern Britain, especially on the drier, eastern side of the country.

Moist American conifer forest

This is confined to the long north–south seaboard of western North America, along the Pacific coast and into the Rocky Mountains, a situation that is to some extent duplicated in northern and western Britain. Several 'moist conifer' species, such as Douglas fir and Sitka spruce, grow faster at first in Britain than in much of their natural range but seem unlikely to live as long because the generally milder winters do not provide a long enough resting period.

North temperate forest

The north temperate forest zone forms a wide band around the northern hemisphere, which includes Britain, where conditions are particularly favourable for good growth. Climatic extremes are rare. Rainfall exceeds the average requirements of most of the zone species in western Britain, and is seldom a limiting factor in the east. Species from the warm southern parts of the zone can be planted only in the mildest areas. If Britain had not been cut off from the rest of Europe 6000 years ago by the English channel it is likely that many more temperate species would have migrated here naturally and been classified as natives.

Southern hemisphere broadleaved mixed forest

The southern broadleaved forest is widely dispersed because land masses in the southern hemisphere are widely separated geographically. Species include *Eucalyptus* and *Nothofagus* (Southern beech). Most of these require warmth and high moisture for satisfactory growth, but some are well adapted to drought. In Britain *Eucalyptus* seed from southern, cooler regions of high elevation are most likely to survive.

Other world-wide forest zones which are recognized include tropical rainforest, savanna woodlands, and tropical deciduous monsoon forest, but none of the trees from these zones will survive out of doors in Britain.

Tree collectors and collections

Food and fodder-producing species and good timber trees have been imported into Britain for many centuries. From before Roman times, elm and chestnut, and probably others were introduced. Information about such introductions is at best sketchy prior to the sixteenth century when published herbals first began to appear. From that time, expeditions were arranged to collect plants, especially to Europe and the Turkish Empire. Soon, exotic fruit trees such as the peach were established in sheltered British gardens and glasshouses. By 1600 the horse chestnut had arrived, and soon after the cedar of Lebanon was introduced. Evelyn's *Sylva* of 1662, the first British forestry textbook, although concentrating mainly on restocking depleted English oak woods, recorded many of the new trees of the time. Landscape architects would at first have nothing to do with 'foreign' trees. Landscapes were contrived to look 'natural' from the point of view of the landed gentry. In the late eighteenth century (according to Loudon), and in complete contrast, only foreign trees were recommended for landscape gardening.

▲

How big can trees get?: a giant cedar on the Welsh border.

Arboretum design developed, particularly in England, during the baroque period, when trees were made to conform to architectural design. Trees were harshly pruned and shaped to formal outlines and often kept in pots to be moved about to fit design criteria. The concept of tree collections for their own sake emerged only when John Tradescant began to discover new species in Virginia. By 1656 several east American trees were established in the Tradescant garden at Lambeth in London. These included false acacia, tulip tree, and swamp cypress. In about 1700 the Bishop of London, Henry Compton, sent English missionaries to convert the American Indians to Christianity. The missionaries were also instructed to collect tree seed, and very soon a fine tree collection was established at the Bishop's Fulham Palace. John Bartram was the first American tree collector; he sent 200 new species to Europe between 1732 and 1760. Some were exchanged for European trees brought to America. By this time the obsession for establishing comprehensive tree collections had taken hold in Britain. Similarly, colonial wealth and world-wide mobility enabled specialist collections to be financed in almost every country, with the exception of China and Japan. The Prince of Wales and Princess Augusta led the way by establishing the Royal Botanic Garden at Kew, where the first curator was appointed in 1759. Exotic tree collections subsequently sprang up everywhere on noble estates. Industrial and inherited wealth were widely available, and often no expense was spared to establish, maintain, and care for trees and arboreta. Collections were usually private and not open to the public, except when the owner had a special exhibit of which to boast.

Secrecy among gardeners was commonplace, and rivalry between the holders of collections was prevalent.

Elsewhere, the Chinese have cultivated trees and gardens for 5000 years, and Japan has a similar horticultural heritage. Both countries were closed completely to western visitors until the early 1700s, and even then only a few collectors were permitted to glimpse their floral riches. The tree of heaven, the Chinese arbor-vitae, and Chinese juniper were sent back to Europe at this time. Visits by Westerners were infrequent and strictly supervised. The British and French made limited inroads into China, while the Dutch concentrated on Japan. In 1776 Carl Thunberg, working for the Dutch East India Company, managed to reach Tokyo. At about this time India, Australia, and South Africa were visited by the British, and the French set off for Peru.

The Pacific coast of North America was investigated by Archibald Menzies, who travelled with the Vancouver expedition of 1792. By 1853 most of the north-west American forest trees had been introduced to Britain. China cautiously reopened its frontiers in 1844; however, the bulk of new introductions from there were not made until after 1899. Japan was visited by John Gould Veitch in 1860 and again in 1890. Charles Sprague Sargent arrived there from America, collect-

▲
Forest tree species are first tested in arboreta and trial plots, such as Speech House Arboretum, Gloucestershire.

ing for the Arnold Arboretum, in 1890. Ernest Wilson moved across from China and began his Japanese work in 1914. Captain Collingwood-Ingram, concentrating on his cherished Japanese flowering cherries, was there in 1927.

The collections that survive in Britain today still contain trees that were established in Victorian times, and there are still fragments of collections and individual specimens which predate that era.

In recent years arboretum inventories have benefited from electronic data processing. Record keeping and the retrieval of information have been transformed by the use of computerized systems. In many establishments much more information is recorded about each tree than ever before, with more certainty than could be contemplated in a traditional planting book or card index. The ultimate aim of botanic gardening is to record the exact origin of every wild exotic tree and to investigate the history of every cultivated plant as far as possible. Regional arboreta have been set up to evaluate and test the performance of the same species in widely different locations. Specialist tree collections are now being established in locations which are likely to suit them best, and records are becoming increasingly compatible and available to other collection holders. Many establishments today use internationally transferable formats so that catalogues can be sent electronically from one collection to another.

Victorian landowners' and gardeners' attitudes of secrecy and competition have nearly gone. There is no longer a need for collections to be comprehensive if this means growing specimens which are obviously unsuitable for a particular climate and soil. There are at present approximately 2500 different trees that will grow in Britain. New plants are still occasionally discovered, and improvement by selection and tree breeding continues. The arboretum remains a cradle for potential forest species.

The growth of the British forest industry

Forestry has its origins in the distant past. People have always needed and used wood, but the emphasis has constantly shifted through the ages, sometimes dramatically. The medieval house builder, for example, required small poles that could be squared or faced up with hand tools. He also needed coppice sticks for 'wattle' to fill the spaces in his timber frames. In the eighteenth century huge oak timbers were in demand for ships. Trees were specially cultivated and selected for 'shipwright' timber with natural curved branches for the ribs of the hull and 'knees' to strengthen the joints. In the first half of the twentieth century emphasis changed again and was placed upon round softwood pitprops and telephone poles. Increasingly through the twentieth century the pulp and paper industries became major consumers of wood. In the future home-produced wood will increasingly need to replace declining imports from natural forests. In 1989, according to overseas trade statistics and FAO (Food and Agriculture Organization of the United Nations) conversion factors, timber consumption in Britain was 50.3 million cubic metres, of which home production amounted to only 6.4 million

▶
Flatford Farmhouse, Suffolk, a medieval timber-framed building.

▲
Crooked-branched English oaks, much in demand by eighteenth- and nineteenth-century shipwrights.

cubic metres. Processed wood products and manufactured boards are easily produced from sustainable home-grown material, but high-quality hardwoods and veneers will be more of a challenge to British growers. The prospects are favourable, for agriculture no longer requires so much quality land, and forestry seems set to move down from the hills and diversify. The age class structure of the British forest estate is widening, the number of young trees is nearly equal to the number of mature trees in our forests. Ancient woodland is now better understood and better protected. Exotic tree species that may be invasive, poisonous, or damaging to the environment are unlikely to be planted or maintained by grant aid. The forestry industry recognizes that tree planting and removal either has a good or bad impact on the landscape, which on a human scale can last more than a lifetime.

British forestry plantations are either operated directly by the state, or privately owned but financed in part through government grants. In 1993 there were 2 162 000 hectares of productive forest, split up as shown in Table 1.

In 1989 over 6.4 million cubic metres of timber was produced, almost 81 per cent of which was conifer wood, because much of the land made available for forestry between 1920 and 1980 was poor, infertile upland suitable only for growing conifers. In addition, the outstanding success of the exotic North American conifers, especially the Sitka spruce, has made their establishment relatively easy and volume production is generally high. In Scotland between 1960 and 1980, 59 per cent of all planting involved conifers, in Wales the proportion was 45 per cent, compared with 3 per cent and less than 2 per cent respectively before 1900.

The forest industry has undergone change throughout its existence. Today the emphasis is on changing back to traditional forest practice to meet various requirements such as wildlife conservation and public recreation and to provide benefits that are not directly concerned with timber and wood products. There is a need to create havens of peace and quiet for relaxation and enjoyment; there is need to enhance the landscape and prevent damage to natural habitats and associated plants and animals. More than ever, though, there is a need for wood, especially

Table 1 Area of productive and other woodland in Forestry Commission and private woodlands. (Taken from *Forestry facts and figures 1992–93.* © Forestry Commission.)

Forestry Commission

Thousands of hectares

	High forest			Total Productive	Other
	Conifers	Broadleaves	Coppice	woodland	woodland
England	174	38	1	213	14
Wales	116	7	0	123	3
Scotland	503	6	0	509	20
Great Britain	793	51	1	845	37

Private woodlands

	High forest			Total Productive	Other
	Conifers	Broadleaves	Coppice	woodland	woodland
England	209	406	37	652	88
Wales	54	57	2	113	9
Scotland	465	88	0	553	65
Great Britain	727	551	39	1317	162

Total Forestry Commission and private woodlands

	High forest			Total Productive	Other
	Conifers	Broadleaves	Coppice	woodland	woodland
England	383	444	38	865	102
Wales	170	64	2	236	12
Scotland	968	94	0	1062	85
Great Britain	1520	602	40	2162	199

Notes: 1. Areas as at 31 March 1993.

2. Forestry Commission figures as at 31 March 1993. Private woodlands figures are based on data obtained from the 1980 Census of Woodlands and adjusted to reflect subsequent changes.

3. Other woodland consists of areas where timber production is not a main objective. It includes areas managed chiefly for amenity and public recreation.

4. Coppice includes coppice with standards.

home-grown wood. This means allowing alternative use of the space in the forest between and under growing trees wherever possible, without losing sight of the ultimate silvicultural objectives of growing the trees. Over 40 000 people in Britain are at present involved in timber and forest-related industries. The most obvious change underway at present is a move to bring woodlands nearer to home. Urban and community forests are springing up on the fringes of towns and cities. Long stretches of motorways and main roads are now well stocked with a wide variety of well-managed trees.

Arboriculture (the cultivation, establishment, and maintenance of trees for ornamental use) is also practised to a very high standard now. Urban park and street trees have probably never been better maintained, or been so numerous. In the countryside, farms have increasing numbers of trees where once there were open fields. Teams of conservation volunteers toil at weekends in coppice woods in much the same way that their ancestors did a thousand years ago. Some are working in exactly the same woods, engaged in active conservation and traditional silvicultural management.

The future

There is little doubt that we shall always need both woodlands and wood products, and in Britain we are short of both. Woods and forests are essential to all the associated ground vegetation and wildlife that depend on them. Today, perhaps more than ever before in Britain, there is greater public awareness of trees and the environment. The Victorian mania for killing and collecting plants, even casually picking wild flowers, has gone out of fashion, and such action is mostly against the law in any case. Vandalism is also less of a problem in most areas now than it was. Forest fires have declined with the eclipse of steam railways and the sparks and hot cinders they produced. There has also been a decline in the number of smokers, and less flammable tree species are generally widely used now. Nevertheless, some threats have tended to increase. Public mobility and curiosity means that there are few havens for sensitive wildlife or places free of human traffic. There are increasing physical pressures on trees and fragile environments are being put at risk through over-visiting, compaction, erosion, and disturbance.

Air pollution may interact with other stresses to cause damage to trees, and carbon dioxide levels are increasing. Trees, which use carbon in their metabolism and structure, might benefit but the consequential warming of the planet may change their distribution patterns and those of their predators. A healthy semi-mature Douglas fir forest can fix 4 tonnes of carbon per hectare per year, and for beech the figure is around 2.5 tonnes. High acidity in streams running out of conifer plantations is another cause for concern. Little of this acidity originates in the forest, much of it coming from industrially polluted air as mist and fog, which is intercepted by the trees and eventually washed away in rainwater.

The considerable amounts of damage that result from storms and climatic extremes cause general dismay. When faced with disastrous storm damage it is difficult to realize that this is all part of a natural cycle of decline and renewal. It may not be economic and often conflicts with our passion for having everything tidy, but there are occasions, particularly in windblown semi-

Primroses, an essential part of the forest floor.
▼

It is often environmentally sensible to leave dead wood lying in the forest, but not where there is a risk of build-up of harmful pathogens.

natural woods, when it is better to do the minimum of clearing up. Broken trees have remarkable powers of recovery and even dead trees have value, especially to invertebrates. Several species grow well from old rootstocks, and natural regeneration by seed is usually rapid. Often it is only after a great tree blows down that new ones have an opportunity to grow.

A simplified key to forest and woodland trees in Britain

This key uses foliage characteristics and some description of the tree itself. It is designed to work with fully grown, ordinary forest and woodland trees; however, with about 2400 types of tree at present growing in Britain, exact identification cannot always be guaranteed. There are brief references to trees not included in the text which will enable readers to pursue their investigations elsewhere. Shrubs cannot be identified with this key.

To use the key you will need a small, freshly cut sample of foliage which includes leaves and a short length of twig (summer foliage for broadleaved trees and deciduous conifers). Notes about the tree such as bark colour, outline, flowers, cones, or any other features which you think might be distinctive will also be helpful.

Start at number 1 on the left of the page. This, and all of the following numbers on the left, will provide you with two or more options. Decide which description best fits your sample and notes. Go to the number or page number indicated on the right after that option. Look up the next corresponding number on the left or turn to the appropriate page.

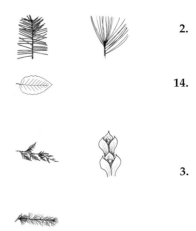

1. A conifer or yew having needle- or scale-like foliage — **2.**

    A broadleaved tree — **14.**

2. Foliage evergreen in flattened sprays, small, overlapping, scale-like leaves covering the shoots completely — **3.**

    Foliage with distinct needles or free-tipped spikes
    (evergreen or deciduous) — **5.**

3. Crushed foliage smells sweet; cones
conventionally shaped, small, leathery with
about eight scales
a *Thuja*; try: western red cedar   p. 37.

The foliage has a bitter smell when crushed   **4.**

4. Pea-sized round cones
a cypress; try: Lawson's cypress   p. 55.
*There are numerous foliage forms of Lawson's cypress
and several similar-looking genera, including true
cypress and juniper.*

Similar to Lawson's cypress but having larger,
woody cones 2 cm across; a very vigorous tree,
often of very great size
Leyland cypress   p. 61.

5. Needles evergreen, short (4–7 mm) in forward
pointing ranks; a very large tree with soft,
red-brown bark
Redwood *Sequoiadendron* (Wellingtonia)   p. 175.
*The Japanese cedar has similar foliage to
Sequoiadendron but the bark is hard
(p. 180 under redwoods).*

Needles conventional, more than 7 mm long   **6.**

6. Deciduous (usually soft and pale green in summer)   **7.**

Evergreen (usually harder and darker green in
summer, needles several years old will often be
found on each shoot)   **8.**

7. Foliage sea-green, needles in tufts on short shoots and singly all round new growth; leathery 3-cm cones with reflexed scale tips, shoot reddish-brown
Japanese larch p. 109.
(Note: evergreen trees with this kind of needle arrangement are cedars.)

Foliage grass-green, arranged as above, cone scales less reflexed, shoot yellowish-brown
European larch p. 103.

The hybrid larch is intermediate between these two.

Other quite common deciduous exotic conifers are dawn redwood and swamp cypress.

The distinctive maidenhair tree (ginkgo) is also a deciduous conifer.

8. Leaves (needles) in pairs, usually more than 8 cm long
Pines pp. 145, 151, 157.
Some other pines have leaves in threes or fives. There are also 14 other species of 'two-needle' pines and many ornamental cultivars but they are not common.

Needles single, mostly less than 6 cm long 9.

9. Needles stiff, straight, often sharp, pointed, and growing on a woody peg which is retained on the shoot permanently when the needles have gone; cones are leathery and drooping
Spruce pp. 181, 187.
There are numerous spruces in cultivation in Britain, and the exact identity depends on minute features.

Needles flexible and not usually sharply pointed 10.

10.   Needles distinctly short (1–2 cm), flattened   **11.**

       Needles mostly over 2 cm long   **12.**

11.   Very large tree with soft red-brown bark
      Coast redwood   **p. 175.**

      Tall, straight tree with hard, dark bark, fine twigs,
      and silver-backed, 1-cm long, flat needles, and
      numerous small cones
      Western hemlock   **p. 91.**

12.   Needles more or less in two flat horizontal ranks   **13.**

      Needles all round the shoot, soft, often blue-green,
      buds are pointed like of those of beech; cones
      have distinctive trident bracts; a large, straight
      tree with often rough, dark, purplish-brown bark
      Douglas fir   **p. 67.**

13.   Tall trees with resinous, often pale grey, stems and
      tiered branches; large, deciduous, vertical cones
      are found at the top
      Silver firs   **pp. 73, 79**
      There are many species and forms of silver fir growing
      in Britain, but the two on pp. 73 and 79 are by far the
      most common.

      Tangled, bushy, or slow-grown standard trees, very
      dark foliage, hard, purplish or red-brown, somewhat
      flaky bark, and soft pink berries in the autumn
      Yew   **p. 205.**

14. Broadleaved trees with opposite leaves or buds   **15.**

Alternate leaves or buds   **16.**

15. Palmate leaves over 7 cm long
a maple of some kind; try: sycamore   **p. 193.**

Palmate leaves, under 7 cm long
Field maple   **p. 121.**
*There are numerous other, ornamental, maples
in cultivation.*

Palmate leaves arranged alternately   **p. 19.**

Pinnate leaves
Ash   **p. 13.**
*Other species of ash occur but they are not common.
Two other genera have opposite pinnate leaves but
they are rarely seen (Phellodendron and Tetradium).*

16. Leaves with an undulating or lobed outline   **17.**

Unlobed leaves (some are toothed)   **20.**

17. Lobed and untoothed
Common and sessile oaks   **pp. 127, 139.**
*Only Sassafras (which is not always lobed), some
other exotic oaks, and Liriodendron also have
untoothed, lobed leaves.*

Lobed and toothed   **18.**

18. Regularly undulating leaves, about 3 times longer
than wide, less than 8 cm long
Sallow willows   p. 199.

Southern beech   p. 25.

More than 8 cm long
Red alder   p. 7.

Irregularly undulating, broad, roundish leaves
(about as long as wide)   **19.**

*There are a large number of different trees growing in
Britain which have vaguely rounded or oval, toothed,
alternate leaves. This is probably the most difficult
group to identify; it is necessary to look closely at the
flowers, fruits, and bark colour, in addition to the
leaves, to make a positive identification. For example,
the common whitebeam, which has pale-backed leaves.*

19. Leaves distinctly hairy and rough. Often a
multi-stem tree
Hazel   p. 85.

Glossy, deep green leaves, often notched at the tip
Common alder   p. 1.

Leaves often felted on the underside, with a short,
obliquely pointed tip
Sallow willows   p. 199.

*There are many alternate toothed and lobed leaved
trees in cultivation, and in the countryside, notably
hawthorn and wild service tree, which could be
confused with a maple.*

20. Simple, virtually untoothed leaves
    Beech   p. 19.
    There are a large number of trees with simple,
    untoothed, alternate, deciduous leaves. In the wild
    Rhamnus frangula is important. In cultivation
    Magnolia, Cydonia, Nyssa, Cornus, and Maclura
    occur frequently.

    Toothed leaves   **21.**

21. Mature leaves, mostly over 15 cm long   **22.**

    Leaves mostly under 15 cm long   **23.**

22. Saw-edged deep green leaves; often large trees
    with vertically (slanted) ridged bark
    Chestnut   p. 49.

    Very fast growing trees, with silver-backed,
    blunt-toothed leaves, buds sticky, and having
    a balsam smell
    Balsam poplar   p. 163.

    Several trees have large toothed leaves; most have
    other very distinct features which are unique and
    aid identification.

23. Heart-shaped leaves
    Small-leaved lime   p. 115.
    Other cordate (heart-shaped) leaves are found on
    Davidia, Morus, and some of the other limes.

    Ovate, lanceolate, or triangular leaves   **24.**

24. Regular, even teeth, ovate shape with glands
     on the stalk
     Wild cherry   p. 43.

     Regular teeth, lanceolate
     Willow (*Salix × rubens*)   p. 199.

     Regular teeth, deltoid
     Black poplar   p. 169.

     *There are a number of different cherries, willows,
     poplars, and other species of trees with these kinds of
     leaves. Specialist botanical descriptions usually need
     to be consulted to distinguish them individually.*

     Irregular, uneven teeth   25.

25. Acutely pointed teeth and parallel leaf veins
     Birch   p. 31.

     Hornbeam   p. 97.
     *Elm leaves can be similar to hornbeam but the leaf
     base is usually unequal.*

     Glandular and uneven, teeth often indistinct
     Sallow willows   p. 199.

1

Common alder

Alnus glutinosa
(L.) GAERTN

Common alder

Alnus glutinosa (L.) GAERTN

Alnus is the Latin name for the tree. *Glutinosa* means sticky and refers to the glutinous secretion on the young foliage. Alder comes from Old English *alor* and *aler*. The Gaelic name is *fearn* reflected in the place-name Ferness in Argyll.

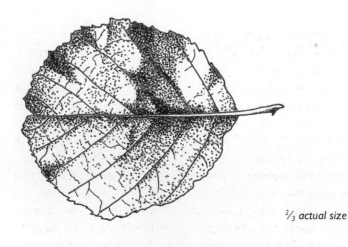

⅔ actual size

◀

This huge alder stem in mid-Wales has missed the coppice worker's axe because of its precarious position.

Origin and distribution

This is a long-standing, pioneer British native species. It reappeared in what is now the British Isles after the last Ice Age, over 7000 years ago. Its natural range includes most types of soil, although it does not thrive on rich dry agricultural land. It may survive up to 490 metres (1600 feet) above sea level. Fossil pollen records are very complete, partly because the tree produces a lot of pollen, and partly because of widespread artificial introduction and use by early man. It is known that alder was the easiest wood to turn on a pole lathe, which is one of man's oldest mechanical devices. It is likely that, in the London area, turners have repeatedly worked their way through alder coppices up and down rivers such as the Lee, Brent, and the Wandle since the Middle Ages. The total distribution of the species includes most of Europe, Russia to Siberia, west Asia, and North Africa. In all of these regions, specimen trees tend to be less common than coppices and pollards. In Britain the tallest recorded height is 32 metres (105 feet) and the largest stem diameter is 132 centimetres (13 feet 7 inches in girth) measured in 1990.

Key features

The alder is a dark, rugged conical, or rounded tree, or a stored coppice clump with several trunks. By 15 years old the purplish-brown bark is already rough and vertically fissured. It becomes progressively more split up into square plates with deep cracks between them, which are reminiscent of old oak stems. The shoot is green at first, and then pale brown with a greyish-purple bloom on the exposed side. Particularly vigorous shoots have prominent orange lenticels. The 7 millimetre ($^{1}/_{4}$ inch) long buds are on distinct 3 millimetre ($^{1}/_{8}$ inch) stalks. They soon turn crimson and purple with a green shaded side. In the spring, new foliage is brownish-red at first, particularly on coppice regrowth, becoming rich green later on. The leaves have a sticky feel, until after mid-summer, which also gives them a glossy appearance. Later in the year they become quite dull. Each has a rounded, irregularly toothed, slightly wavy outline, and sometimes a notched tip. The seven pairs of nearly white parallel veins

▲

Female fruits of the alder shed their seed, and then turn black and stay on the tree all winter.

under the leaf are distinctive. The leaf margin is somewhat extended down the stalk, which is seldom more than one-third of the length of the leaf. It is also pale with tiny brown specks along its length. Foliage stays late on the tree in the autumn, but remains a dull green colour until it falls. Male catkins appear 7 months before they open, remaining in tightly closed bunches of three all through the winter and eventually opening in March. Female catkins are borne as short, erect clusters. They ripen in summer to globular strobiles up to 15 millimetres ($\frac{5}{8}$ inch) long. After shedding their seed, these turn nearly black, become very hard, and may hang on the tree for another year. The tiny seeds have a pair of buoyant wings, which enable them to fly on the wind or float on water.

Growth conditions

Soils that are waterlogged, but not too acid, are suitable for the alder. It likes the proximity of water, growing well on marshes, lakesides, and fens, but not acid bogs. It requires little free oxygen in the soil, and appears to be adequately supplied by oxygenated fresh water. The alder's habitat is shared with shrubby grey sallow, buckthorn, alder buckthorn, and guelder rose, and ground vegetation including yellow flag, nettle, meadowsweet, hemp agrimony, and sedges. All of these, and the alder, are good food plants for wildlife. The alder itself supports over 90 organisms. The tree will not stand excessive exposure to the wind, although it is usually stunted and not killed by wind.

Propagation and management

Propagation from seed is easy, and there can hardly be a more tolerant seed, but it does not remain viable for long. In semi-natural conditions it will germinate in almost pure pebbles or sand, and will even grow in the masonry of canal and dock sides. In this respect, in Britain only *Buddleia* can compare with it. Planting sites can also be nearly 'soil free'. Rocks, gravel, bricks, and clay can contribute up to 80 per cent of the rooting medium on reclamation sites on which the alder will grow. Part of the tree's independence of good soil is its unusual capacity to obtain nitrogen from the atmosphere using root nodules that are inhabited by nitrogen-fixing, bacteria-like organisms. Tests have shown that young plants artificially provided with these organisms at an early stage establish rather better than untreated plants on ground that has not supported alder before. It is said that, given rich soil and good drainage, the plant will

◄
The alder requires very little soil in which to grow, and often invades brickwork as here at Lydney dock.

◄
This alder stand at the head of Loch Fyne in western Scotland must have its roots in brackish water.

grow into a large bush and never make a tree. Traditionally alder was coppiced on a 10-year rotation for turnery and somewhat longer for making clogs. Poles were scored or 'scorched' by chipping off patches of bark with an axe to enable the wood to season more quickly and evenly when stacked outside. Larger stems were grown for plywood or rotary peeling. Traditional management of alder woods has declined and many old coppices have been grubbed out.

Timber quality and uses

The timber is of medium density but is relatively soft, porous, diffuse, and not durable unless kept permanently under water. Green wood when first cut is pale coloured, but in a few hours the cross-cut ends turn bright orange. Once seasoned the wood is uniformly pinkish-brown with a radial sheen. 'Scots mahogany' is alder wood that has been immersed in a peat bog

Alder wood is soft and porous, but is very durable if kept wet.

directly after felling until the reddish stain becomes permanent. Swellings and stem burrs have always been in demand for inlay work on fine furniture. Veneers, pulp, and plywood have been the most recent uses for alder timber. Traditionally the wood was associated with use in or near water, for which it was well suited: it was made into pumps, piles, sluices, troughs, and faggots that were placed in drain bottoms and covered with stones and earth for field drainage; small boats and punts were built with it more easily than with elm; river banks were strengthened with alder posts, which often rooted and provided a binding root mat and bushy cover as well as a 'rain' of insects for the chub that lay below. Strangely, for a wetland tree that takes a long time to dry out, charcoal was made from it and used for gunpowder. The bark was used for tanning, and when soaked in company with iron produced black ink. The most notable alder industry of the past, however, was clog making. Until the early twentieth century clog sole cutters worked in gangs moving from coppice to coppice as the wood became ripe for cutting. Northern England and the Welsh valleys were their stronghold. Using a cross-cut saw and a 'stock knife' (a fixed blade on a bench) rough shoe soles were cut out and stacked in an open pile to dry. Once seasoned, and considerably lighter, they were carted off to a factory for finishing. Today, the alder is valued as a living tree more than for its timber, and there are various ornamental forms including golden-leaved and cut-leaved cultivars. On estuaries, this is one of the last woody plants to be thwarted by salinity. The alder can be found growing out between brackish creeks almost to the limit of terrestrial vegetation. It may be employed again for flood control, which it does by fixing soil with its roots and catching silt and debris with its shoots. Eventually, on freshwater sites, alder 'carr' is formed, which is an important natural transition between marshland and scrub woodland.

2

Red alder

Alnus rubra

BONG

Red alder

Alnus rubra BONG

Alnus is the Latin name for the alder tree, and *rubra* comes from the Latin word for red, *ruber*. The common name is derived from this; many different parts of the tree, notably the twigs and buds, are red.

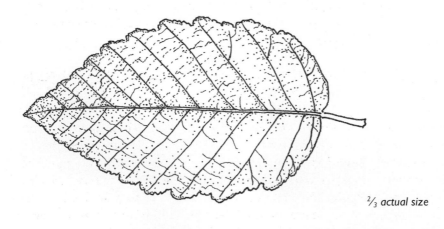

²/₃ *actual size*

Origin and distribution

Red, or Oregon alder, comes from the western side of North America, from Alaska to California and Idaho, and was introduced into Britain before 1880. It does not live very long in this country, and the oldest surviving tree was probably planted after 1930. It has become a widely planted woodland tree relatively recently, partly thanks to its ability to survive on very poor soils by fixing nitrogen through a bacteria-like micro-organism of the genus *Frankia*, which forms nodules on the roots. It has proved to be an ideal tree to rapidly cover spoil heaps and reformed land. It is wind-firm and may be used to nurse more valuable species, providing shelter, stability, nutrient-rich leaf litter, and nitrogen-enriched soil. It has been used throughout the British Isles on a limited scale and thrives on both extremely alkaline farmland and quite acid sites. The tallest specimens known are at Threave Castle in Dumfries and Galloway at 23 metres (75 feet), and 58 centimetres diameter (6 feet in girth) at Kew measured in 1989. In addition to red alder, trials of grey alder (*Alnus incana*) are in progress, mostly on reclaimed tips and sand dunes. Italian alder (*Alnus cordata*), which grows extremely fast in

▲
Red alder stems seldom grow very large.

the south, is being planted increasingly as a street tree and in small plantations, especially on lime-rich, dryish, alkaline soils; it does not coppice like the grey alder. The tallest specimen in Britain, at Westonbirt, is 34 metres (111 feet).

Key features

Red alder in its natural habitat, particularly in Washington State and Oregon, has silvery-white bark, rather like the American aspen and birch. In Britain, however, its stem is always a dull grey or pinkish-grey, and smooth for most of the tree's life. The crown is narrow for ten to twenty years, eventually becoming broad and somewhat angular and untidy with age. The shoots are green at first, but soon turn a rich dark red and become glabrous. The buds are also red, sometimes with a green shaded side. The transition from red twigs to pale trunk spans several years, during which time most of the crown in winter has a red-brown to grey-brown appearance. The leaves are 12 centimetres long and have 10–12 parallel veins which are promi-

▲
The leaves are large for an alder, reaching 12 centimetres long by 6 centimetres wide.

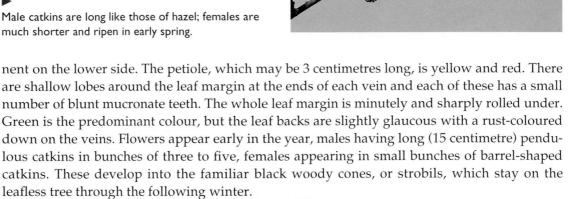

▶
Male catkins are long like those of hazel; females are much shorter and ripen in early spring.

nent on the lower side. The petiole, which may be 3 centimetres long, is yellow and red. There are shallow lobes around the leaf margin at the ends of each vein and each of these has a small number of blunt mucronate teeth. The whole leaf margin is minutely and sharply rolled under. Green is the predominant colour, but the leaf backs are slightly glaucous with a rust-coloured down on the veins. Flowers appear early in the year, males having long (15 centimetre) pendulous catkins in bunches of three to five, females appearing in small bunches of barrel-shaped catkins. These develop into the familiar black woody cones, or strobils, which stay on the leafless tree through the following winter.

Growth conditions

This tree is often used, and is most valuable, as a nurse to other species on difficult sites. In nature, it is a pioneer of bare ground which it will colonize very quickly from seed. It is light-demanding and invasive, and is well able to outgrow most weed species, particularly if the trees and the weeds start out together on a bare site. On surface water gley soils and heavy clays that are slow to dry out, red alder will cause fissures to develop which, once opened, will not completely close again the following winter. The effect of this is to increase the depth of rootability through oxygenation and vertical drainage of the soil: stability and nutrition are

▲
Young red alder woods grow very rapidly in wet conditions.

improved, ground water becomes sweet and usable, and the drier soil warms up earlier in the growing season. Early growth can be quite exceptional, annual shoots reaching as much as 1.5 metres (5 feet) in length for about ten years. A tree at Westonbirt reached 18 metres (59 feet) in 12 years. A small plot on Wykeham Moor in Yorkshire planted at 2-metre spacing (6 feet 6 inches) on a felled pine site was ready for thinning out and produced a crop of small poles after only 5 years.

Propagation and management

Seeds of red alder are minute, and 1 kilogram of them may contain 1.3 million. At least half a million seedlings after one growing year may be expected. Plant type may be critical on some sites. In trials in north Devon in 1962, tall thin plants grown up to 50 centimetres (20 inches) with nitrogen in the nursery failed completely outside, whereas shorter stout plants survived weed competition and spring frost on the same site. As a nurse species on wet sites red alder is most valuable, but care must be taken that it does not grow so fast that it suppresses the species being nursed. For example, Sitka spruce may be defoliated or deformed by adjacent alder branches. Although forest trees will often eventually kill the alder nurse, any surviving alders can be coppiced for reuse with a subsequent crop.

Red alder wood is an important commercial hardwood in America.

Timber quality and uses

The wood of red alder is of medium density and fine in texture, but lacks much distinctive figure or grain. It dries well, turning a pale brown colour, and does not shrink as much as the common European alder. It takes a good finish when polished, and takes paint or varnish very well. In America high-quality logs make good veneers for plywood. It is the most common commercial hardwood in parts of its natural range, where it is used for furniture, various small industries, and craftwork. Children's toys and turned items are made from it, particularly tool handles. Very little alder wood is produced in Britain for such purposes. Rather, the tree is valued as a living entity to disguise the problems of industrial land use and tipping, to line city streets and ring-roads, and to aid the establishment and growth of more valuable trees.

Ash

Fraxinus excelsior L.

The genus name *Fraxinus* is from the old Latin name for the tree, believed to be derived from the Greek *phrasso* to fence—a common use for the living trees, or their wood. *Excelsior* in Latin means literally 'higher' referring to the great height of the tree. Ash comes from *Aesc*, the Anglo-Saxon word for spear.

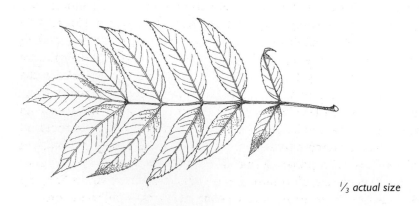

⅓ actual size

This pre-1700 tree in Gloucestershire has now been pollarded.

Origin and distribution

Ash is a native species that has spread naturally over the whole of the British Isles. It will not thrive on the most exposed mountain areas, but is found growing at elevations of 370 metres (1200 feet). The natural range of the species extends eastward across Europe to the Caucasus. Because of its great economic value in agriculture and war as a strong pole or stave, it is thought likely that ash was in fact introduced by early man ahead of its natural post-glacial spread. It is one of the largest deciduous trees in Europe, frequently reaching 40 metres (131 feet) in height. In Britain the tallest tree on record is 38 metres (125 feet) and the largest breast height diameter in 1988 was 281 centimetres (29 feet in girth).

Key features

Ash trees are light in colour of branch, stem, and foliage. They are tall and graceful with widespread billowing crowns, huge arching branches, and angular, twisted, stout twigs. These may be upright on young and middle-aged trees, and pendulous on old specimens. In winter ash is easily recognized by its pale grey twigs, which tend to clatter together in high winds, and its large, black, velvety buds in terminal clusters and opposite pairs. Ash shows marked clonal variation and young shoots may range in colour from grey-green to purplish-brown on different individual plants. This feature shows up well on young coppice. The pinnate leaves are pale green, with between 6 and 12 pairs of lanceolate to ovate leaflets and a single slightly larger leaflet at the end. The margins are shallowly toothed. The rachis or leaf stalk is yellowish or sometimes tinted with purple. In the autumn the foliage breaks up and falls without a great show of autumn colour, although some individual trees may turn dusty yellow in late October. Dark maroon flower buds appear in April well before the leaves expand. These can be male, female, or hermaphrodite produced all on the same tree, or on distinct male or female trees. The male stamens are deep red-black at first, becoming paler and then yellow with pollen at maturity, while the female flowers start very small and insignificant but quickly develop into single-winged green seeds in hanging bunches. In the autumn these seeds or 'keys' turn a straw colour. They may fall to the ground quickly or hang on the tree through the winter months, depending on the weather.

▲
Ash foliage needs good light to thrive.

▲
The flowers of the ash are small and difficult to see. In spring they turn from black to brick red before shedding yellow pollen.

▲
Seeds hang in familiar bunches of 'keys'.

Growth conditions

The best ash sites are on rich lowland, fairly alkaline, moist farmland soils. The tree is hardy enough to survive almost anywhere in Britain, but does not thrive on poor quality land. Clay soil will support good specimens but moist limestone soil is better. Ideal ash sites are most often used for agriculture, so many of the best trees are found along farm hedgerows. The ash regenerates freely and seedling survival is often very good. A few young saplings will compete for a time in woodlands with oak, but cannot ultimately survive under beech. Mature woods of pure ash are not common, as competition between individuals for light is fierce. After the first few years, if the trees do not have enough light, proper development and growth will become stunted. Even if the competition is removed, they rarely regain their lost momentum. Grazing animals, especially rabbits, often eliminate unprotected seedlings. Ash trees are late coming into leaf and so usually escape late spring frost damage. The trees have great strength and seldom snap or blow down in the wind.

◄
The ash is a light-demanding tree
that thrives in hedgerow isolation.

Propagation and management

The ash is usually grown from seed although it is very easy to graft. Seed is produced after about 25 years of age, and heavy crops tend to occur at intervals of 3–5 years. If collected green in August and sown immediately, dormancy may be avoided. Seed collected from October to November will require stratification and will not germinate for two years. About 2500 seedlings may be expected from 1 kilogram of seed. Good management of ash depends upon a good choice of site, protection from damage, and ensuring at all times a sufficient amount of light to maintain even growth.

Timber quality and uses

When ash grows fast it forms broad, evenly spaced annual rings and becomes exceptionally strong. It is one of our toughest native timbers, withstanding pressure, shock, and splintering because of its flexibility. It has long been used for making tool handles of every kind and for sports equipment and coach building. Artificial materials such as nylon, steel, or plastics, and improved industrial technology, have only recently pushed ash out of common use in these areas. It does not have the durability of oak or elm in contact with wet soil, but little can equal it for

▲
This stem has even growth but it also has a 'black heart', a black stained central core of wood.

strength and elasticity in dry places. High-quality furnishings, particularly fittings in hotels, schools, shops, and public buildings are increasingly made of ash, where it is usually seen with a clear, pale, glossy finish. It is used to great effect in laminated work and preformed plywood structures.

4

Beech

Fagus sylvatica

L.

Beech

Fagus sylvatica L.

Fagus was the name given to this tree by the Romans, and it may have been taken from the Greek *phagein* meaning to eat, referring to the beech nuts or mast. *Sylvatica* refers to woodland. The English name 'beech' is related to the early German and Scandinavian *buche* and *bok*. The word 'book' in English has the same origin. Norse writing was often done on thin sheets of beechwood.

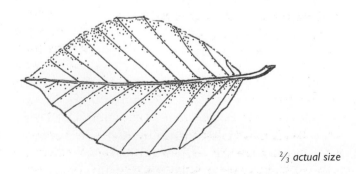

²/₃ actual size

▲
Beech woods tend to have a clean floor.

Origin and distribution

The beech tree is certainly native to southern England and south-east Wales and has been introduced northwards to the whole of Britain from an early date. It thrives on the limestone soils of the south and east, provided there is an acid surface layer, where it is second only to oak in importance as a broadleaved woodland tree. The natural distribution includes the whole of central Europe from Denmark to the Pyrenees, although it is absent from most of the Mediterranean coastal strip, and only just extends into Turkey and Russia. There are small isolated natural woods in Sweden, Norway, Corsica, and Sicily. In Britain the best-quality trees are found in the Cotswold Hills, the Chilterns, and in Sussex. However, the largest specimens are in Scotland, the tallest on record reaching 46 metres (151 feet) and the greatest stem diameter known being 231 centimetres (23 feet 10 inches in girth) at breast height (this specimen has recently fallen).

Key features

This huge, fast-growing, deciduous tree usually has smooth silvery grey bark which is retained throughout its life and into old age. The slightly zig-zag shoots are, like the leaves, at first covered in silky hairs but soon become quite glabrous. Each young gossamer-fringed leaf retains a long brown stipule at the base of its reddish stalk until June. The leaf blades are 7 centimetres (3 inches) long, oval to ovate with a pointed tip and slightly rounded base. Anything from five to nine straight parallel veins run out from the equally straight mid-rib to the undulating, but entire (untoothed) margins. Beech flowers open in May, at the same time as the leaves. Male flowers hang down in tiny bunches on slender stalks. In late May and early June drifts of discarded male flowers and leaf bracts litter the forest floor with pale brown debris. The cup-like female flowers stand upright on short stalks, each having four bracts covered in fine silky hair. These develop into the familiar, hard, spiky, four-sided cupules, each of which usually contains two triangular seeds which become glossy rich brown beech nuts. Good seed years, called mast years, occur only at intervals of three to five years. In such seasons enough seed is produced to satisfy the numerous creatures that feed on beech seed, and at the same time provide some to grow into new trees. Autumn foliage colours are outstanding, the leaves changing to every shade of yellow, gold, and finally lustrous brown. Some trees retain a few

dead leaves through the winter on the lower branches. Olive brown winter twigs support prominent, slender, sharply pointed buds, which are spaced alternately and stand out slightly from the shoot. From some distance away beech trees have a distinctive rich brown colour in winter resulting from a combination of twigs, buds, and dead leaves. Old beech trees may decline very quickly indeed if diseased or under severe stress. Unlike oak, the beech will not 'hang on' for generations as a half-dead hulk. Trees over 150 years old are unusual, although specimens up to 250 years old have been known. The best copper beeches are selections made by nurserymen originally from naturally occurring purple-leaved forms. Spontaneous purple-leaved seedlings still appear regularly, their green leaf colour being more or less masked by red pigments.

◄

This 170-year-old tree at Westonbirt Arboretum is over one metre in diameter.

Early summer flowers appear with the leaves.
▼

Autumn brings a riot of colour.
▼

Growth conditions

The beech is a shade-tolerant tree able to regenerate under the shade of other trees but seldom under other beeches. Beech leaf litter tends to acidify forest soils over a long period which may also inhibit natural regeneration. It is widely naturalized, however, and may seed freely on many different types of site. It cannot withstand late spring frosts and young plants are scorched by strong sunlight. Once established, beech forests will shade out all other vegetation, even the bluebells and brambles with which they are associated. Severe gales occasionally cause windthrow if deep rooting has been impeded or roots are damaged by fungi or excavation, and old trees can suffer severe branch and even stem damage. The beech is prone to producing weak branch forks which eventually tear apart. The thin bark is damaged by squirrels and rabbits in the spring and early summer—branches or whole trees girdled by these animals will die. A serious disease of beech is the parasitic beech bark disease fungus, *Nectria coccinea*. Initial infestation by the felted beech coccus, a minute sap sucking insect, allows this fungus to penetrate the bark and spread into the live tissue beneath.

▲
Mature Gloucestershire beech woods.

Propagation and management

Propagation of beech is not difficult, provided that young emerging seedlings can be protected from direct sunlight, late frosts, and mice. The provenance, or origin, of seed is important, and only tested and approved sources should be used if timber production is the first objective. Seed from Forêt de Soigne in Belgium is considered to be one of the very best. A kilogram of seed contains about 4600 individuals which may be expected, if grown under shade, to produce 3000 usable seedlings. Young plants in the wood will survive under dense vegetation and overhead shade. Closely grown trees are usually fairly knot-free and straight, but in an even-aged wood dominant trees are slow to assert themselves. Thinning out must not be delayed or the trunks may become elongated and weak. A wide range of age classes will coexist very well together in beech woods.

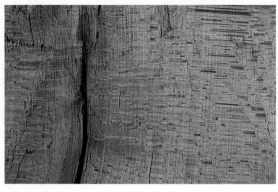

▲
The finely figured timber is highly prized for furniture and internal joinery.

Timber quality and uses

The timber is heavy and strong, pale pinkish-brown in colour with numerous, small, radial flecks (medullary rays) of warm brown. The division between heartwood and sapwood is indistinct. It is excellent for all kinds of furniture, turnery, laminations, and manufactured goods. It is seldom used in large quantities for construction work, or in an untreated state, because it is susceptible to insect attack. Beech wood is not durable outside without preservative treatment. It is clean and nearly odourless, so is suitable for uses associated with food, and it makes excellent kitchen fittings and children's toys. Veneers of beech are incorporated into some high-quality plywood. After treatment with steam and heat, the wood is ideal for 'bent-work', particularly the curved backs of chairs. In the timber trade, many unrelated hardwoods are called 'beech', and they come in all colours—European beech is known as 'red beech'. As a living tree beech provides good shelter and is highly valued for amenity. Unfortunately, at all ages its thin bark is vulnerable to vandalism. It is a perfect garden hedge plant, and if clipped in summer will retain an even coverage of dead leaves through the following winter, truly a permanent, all-year screen that changes with the seasons.

5

Southern beech

Nothofagus nervosa

(PHIL.) DIM. AND MILL. (*PROCERA* OERST.)
AND *Nothofagus obliqua* (MIRBEL) BLUME

Southern beech

Nothofagus nervosa (PHIL.) DIM. AND MILL. (*PROCERA* OERST.)
AND

Nothofagus obliqua (MIRBEL) BLUME

Notho means false, and *Fagus* is the generic name for beech. *Nervosa* is from the Latin, a reference to the conspicuous nerves or veins on the leaf. *Procera* means 'tall' from the Latin *procerus*. *Obliqua* is Italian for 'turn aside', alluding to the unequal-sided leaf bases. The common name for this genus refers to beech-like trees from the southern hemisphere.

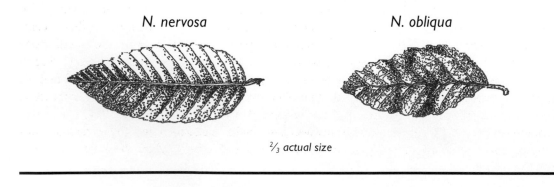

N. nervosa   N. obliqua

²⁄₃ *actual size*

▲
Large stems are becoming common as original plantings mature.

▲
This *N. obliqua* is an original tree photographed when it was 85 years of age.

Origin and distribution

The 40 or so beeches of the southern hemisphere are mainly evergreen with two notable decid-uous exceptions from Chile and Argentina which have been found to grow well in parts of lowland Britain. *Nothofagus nervosa* is a native of the Andes and a coastal strip in Chile, and there are a few stands in Argentina. It was introduced to Britain by Arthur Balfour of Dawyck who made plants available from 1914 onwards. *Nothofagus obliqua* has a natural range which extends further to the north, to a climatic zone somewhat similar to the Mediterranean area. In the past, whole extensive native forests have been felled to make way for grazing. It is possible that William Lobb introduced the tree into Britain in 1849, but this is by no means certain. The earliest definite introduction was in 1902 by Henry Elwes. Balfour also introduced this species along with *N. nervosa* soon after. Original trees of both species survive in Britain, so the ultimate age and size has yet to be reached. In west Sussex a *N. obliqua* has reached 36 metres (118 feet) but it fell in the 1987 storm. In Cumbria a *N. nervosa* has been recorded at 30 metres (98 feet) (it has reached 32 metres in Ireland). The largest stem diameter of *N. nervosa* in 1988 was 131 centimetres (13 feet 6 inches in girth) at Brodick and the largest *N. obliqua* anywhere in 1992 was under 100 centimetres (10 feet 4 inches girth). *N. alpina* was described at the same time as *N. nervosa* and was then thought to be a small-leaved form. It is now considered to be a hybrid between these two species (*N.* × *alpina*).

Key features

These two trees are large and usually fast-growing at first. The majority of specimens, however (except in the extreme west of Wales and south-west England), tend to slow down with age, particularly in height growth. They are well suited to either forest or open woodland environments. Both species have pale, fairly smooth beech-like bark with a tendency to flake or produce long vertical fissures on the oldest trunks. *N. nervosa* has distinctly larger leaves (9 centimetres, 3½ inches), larger in fact than any other deciduous southern beech. The leaves resemble hornbeam leaves, with 18 conspicuous parallel pairs of veins and sharply toothed margins, and are alternately arranged on new shoots in a way that is reminiscent of some elms. *N. obliqua* has much smaller leaves, with only about nine pairs of veins, which are also arranged in two rows, and an oblique base like the common elm. The leaves of both species are dark green above and paler beneath. In the autumn they colour well and are retained for a long time. The shoots are green, then dark brown and warty. Male flowers with up to 40 stamens are produced singly in the leaf axils, and female flowers develop into fruits containing three nutlets, the centre one being flattened and the other two being triangular. They are contained in a softly spiked and fringed glandular husk. Winter buds are chestnut brown and sharply pointed, rather like hornbeam buds.

N. nervosa leaves.

▲
N. obliqua leaves.

Growth conditions

Although both of these species are regarded as hardy in the British Isles, even in southern England they may occasionally be affected by cold weather, perhaps because they grow actively late in the year. This is likely to cause dieback of young shoots, possibly resulting in multiple stem development. If cambium is killed by cold, stem cankers can develop and become a breeding ground for disease. Complete girdling will kill young trees. Late spring frosts may damage young foliage, but selection of southerly seed origins will go some way to avoiding this. *N. nervosa* should only be planted on mild, moist, preferably sloping sites. *N. obliqua* of southern origin will grow on slightly colder ground. Neither species is suitable for exposed, waterlogged, shallow, very chalky, or infertile sites, although they will grow on soils of low nutrient status where many other broadleaves would fail. Deep acid peats are quite unsuitable. Broadly speaking, of the two, *N. nervosa* is best suited to the milder parts of western Britain, and *N. obliqua* to the drier parts of the south-east. Insects have not become a serious pest on these trees in Britain, but fungi such as *Fomes annosus* and *Phytophthora* have killed some specimens. The light foliage of *Nothofagus* allows a rich woodland ground flora to develop which can be similar to that of an oak wood. In addition, both species are found to support good assemblages of insect life, which is surprising for such a very recent introduction.

Propagation and management

Seed is the most effective means of propagation, which home-grown trees begin to produce from about the age of 23 years. There are about 116 000 seeds to one kilogram, of which 25 000 should survive as plants after one year. Seedlings should remain in the nursery for two years before planting out. Vegetative propagation is possible using summer cuttings using mist and heat under glass. Roots are very susceptible to drying out, and must never be left uncovered before planting. Plantations are few and far between, so little is known about their management. It does seem, however, that heavy thinning is required, particularly of the faster growing *N. nervosa*. Failure to do this will reduce

▲
N. nervosa grows best in wet western areas, West Wales being particularly favoured.

Seeds occur in small, woody glandular husks.
▼

growth of weaker individuals to nothing and diminish crowns completely. The considerable annual growth of some stands is no doubt due to the very long growing period, from March to September. This, however, has the disadvantage of causing the trees to suffer from cold conditions at the start and end of each growing season. Natural regeneration is frequent on favourable sites and cut stumps coppice freely.

Timber quality and uses

N. obliqua yields durable reddish timber. The Chilean name 'roble', frequently used in Britain, is Spanish for 'oak'. In its native countries it is used for joinery, furniture, and shipbuilding. The sapwood is soft and white. *N. nervosa* grown in South America has timber similar to beech in Europe. In Britain the area of *Nothofagus* woodland is still very small indeed, and much of this was only planted after 1956. Most of the thinnings cut down so far consist largely of soft sapwood, and the real value of the timber has yet to be seen. Investigations continue into different seed origins of both species to find plants that are both reliably hardy and productive.

▲
The timber is not unlike European beech.

6

Silver birch

Betula pendula

ROTH

Silver birch

Betula pendula ROTH

Betula is from the Celtic *Betu* meaning 'tree-like'. *Pendula* refers to the drooping young branches. Silver refers to the colour of the young bark, and birch is from the Anglo-Saxon *'birce'* meaning 'a tree'.

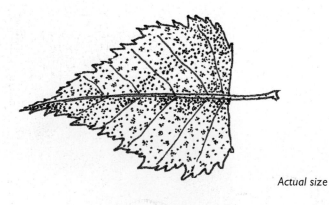

Actual size

Origin and distribution

The silver birch is a very hardy tree, native throughout the British Isles. Its natural range extends right across Europe to Asia Minor and northwards, well into the sub-Arctic. It spread from continental Europe into Britain immediately following the retreating glacial ice sheets some 12 000 years ago. A large number of species of insects and mites are associated with it, only surpassed by the oak and willow. The birch has adapted to local photoperiod and temperature conditions in Britain, so dates of flushing and leaf fall vary from one area to another. Best results from planted trees are likely to be achieved if seed is collected from good local sources and not imported from a great distance. Although the largest British specimens ever recorded have been found in the south of England, high-quality specimens are found in every region. The largest stem diameter on record measured in 1987 was 133 centimetres (13 feet 8 inches in girth) at Leith Hill in Surrey. The tallest tree known in 1989 was at Birse in Scotland, standing at 30 metres (98 feet).

The birch frequently invades heathlands such as here in Surrey.
▼

Key features

The birch is a graceful tree, with light foliage and an open head of branches. It is best known for its distinctive peeling bark, which is red, pink, and brown at first, then silvery white for many years. With advancing age dark, nearly black, patches, punctuate the white, gradually building up from the base of the tree. The whip-like branches and resinous, thin, warty twigs tend to droop at their tips. The conical winter buds are very small and somewhat sticky. The leaves, which turn a glorious golden colour before falling, are simple, alternate, and distinctly toothed; they are ovate, sometimes truncate at the base. There are deciduous stipules adjoining each leaf stalk usually remaining until mid-summer. Separate male and female flowers are borne on the same tree and, as no scent or nectar is produced, pollination is dependent upon the wind. Groups of developing male catkins hang on trees throughout the cold winter months then open out to form lax pendulous tassels in the spring. Female catkins (*strobiles*) containing clusters of tiny flowers are shorter and more ample than the male and do not droop. Single-seeded winged nuts containing the true seed ripen in late summer. The *strobiles* then slowly disintegrate, sending seed and catkin debris away on the wind together. Seeds carry little nourishment with them, so germination is a matter of some urgency. Birches tend to be short-lived except in mountainous areas.

▲
Birch leaves are more or less triangular and colour well in the autumn (yellow to old gold).

▲
Male catkins in spring give a fine but brief show.

Growth conditions

Silver birch is characteristic of sandy heaths and mountainsides, but it will tolerate many other sites in Britain. It tends to be replaced by the downy birch, *Betula pubescens* Ehrh., on very wet soils. Although there are forms of both species that show characteristics of the other, genuine hybrids are rare and usually infertile. The tree is cold-tolerant, although imported less robust plants may be damaged by frost when moved long distances from south to north, or west to east. Trees are generally wind-firm because of their spreading roots and light superstructure. Those that do blow down are often on ground that is mechanically so unsound that it cannot support them. Full light is required for good development, although some shade is tolerated by very young plants. Abundant seed is produced, which quickly germinates on any bare or disturbed ground. Natural regeneration of even aged specimens is often seen following heath fires or earth-moving operations. One of the many fungi associated with birch woodland is the poisonous Fly agaric which lives in a symbiotic association with the tree roots.

The downy or white birch, *B. pubescens*, is evenly and widely spread over the British Isles, where it is normally associated with damp areas, especially the highland glens of Scotland. It rarely forms woods of large trees, preferring to remain a thicket of often numerous small stems. Downy and silver birches can be easily identified by their young shoots, which have soft down on the former and rough warts on the latter.

▲
Birchwoods on peat in Humberside were often a consequence of fires caused by railways and industry.

Propagation and management

Seed collected in early autumn should be sown in containers for the best results. A kilogram may contain 1.9 million seeds, 150 000 of which should produce usable seedlings. Young plants resent any root disturbance and care should be taken to avoid this when they are planted out. Intensive ground preparation is not usually necessary before planting, and young trees usually grow very well in plastic treeshelters. Natural regeneration usually occurs frequently; this, together with coppice growth from healthy cut stumps on suitable sites, will provide an infinite number of crop rotations without any need for artificial planting. If at any stage live branch pruning is required, it should be done during the summer months. Branches cut off in winter will cause the tree to 'bleed' the following March when the sap surges upward.

▲
The fruits break up in early winter, spreading tiny seeds and scales far and wide.

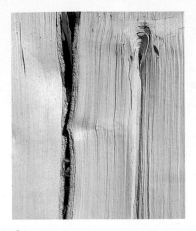

▲
Birch timber is a beautiful, pale golden colour combined with reasonable strength.

Timber quality and uses

There are numerous traditional uses for birchwood, bark, and even twigs. Untreated timber is not suitable for permanent outside use but it is often used for furniture and small turned items such as tool handles and broomheads. Treated with a clear finish, the wood looks golden brown with a silvery medullary sheen. Timber grown in northern Britain sometimes contains random red-brown pith flecks caused by the *Agromyza* fly. This damage does not appear to weaken the wood, but alters dramatically its visual appearance. In Europe and Scandinavia birchwood is highly prized and widely used for decorative and functional household items. Occasionally foresters make use of the light shade under heavily thinned-out birch woods to establish crops of alternative shade-tolerant trees.

Western red cedar

Thuja plicata

D. DON

Western red cedar

Thuja plicata D. DON

Thuja is from the Greek *thyon* or *thia*, a tree that produced resin or incense used in religous ceremonies. *Plicata* means 'folded into plaits', referring to the foliage. Red cedar is really the commercial timber name—the tree is in fact a cypress. In former times this tree was called *arbor-vitae*, 'the tree of life', due to its evergreen foliage.

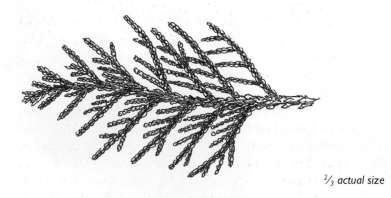

²⁄₃ *actual size*

◀
A tall tree is produced eventually
but the lower stem tends to
be fluted.

Origin and distribution

This is a North American Pacific coast tree which has a huge north–south range from south-east Alaska to California, with a second inland distribution from British Columbia to Montana and Idaho. It reaches a majestic 50 metres (164 feet) in height on favoured sites in its natural habitat. It was introduced to Britain by Thomas Lobb in 1853, although it is possible that John Jeffrey brought seed back from America to Scotland two years earlier. The late introduction date is surprising because the plant was described as early as 1794 by David Don, and Archibald Menzies was known to have collected seed in 1795. The oldest surviving plantation of the species is at Benmore in Strathclyde. The tree is not widely planted as a forest tree in Britain but is found in many places as an ornamental specimen or hedging plant. The tallest tree known in 1985 was 46 metres (151 feet) and the largest stem diameter in 1987 was 193 centimetres (20 feet in girth).

Key features

The crown of the tree is columnar and the foliage is both billowing and weeping. Old isolated specimen trees have branches that often weep to the ground and layer freely, ultimately making a ring of new stems. The fibrous, vertically ridged, and slightly peeling bark is fairly soft and reddish-brown. In old age the base of the tree becomes fluted. The shoot is green for a year or two, and then turns bright orange-brown. The glossy green foliage, which obscures the whole shoot for a time, is held in flat or drooping sprays. The individual leaves appear as tightly packed, very small waxy scales with a central translucent gland like a pin-hole which can be seen against the light. The underside has a pair of pale stomatal bands. The foliage is sweetly aromatic, reminiscent of crushed orange peel. Male flowers are small and globular on the shoot tips and females are flask-shaped and erect. The cones are leathery and small with up to ten soft hinged scales, only five or six of which are fertile and develop properly, each producing two tiny winged seeds. The easiest way to distinguish between this species and the very similar Lawson's cypress (Chapter 10) is by the smell of the crushed foliage and the type of cones.

Growth conditions

Anaerobic soil conditions exist in parts of this tree's natural habitat, which it is well able to tolerate. It is also very shade-tolerant and will survive for many years under the shade of other trees. It prefers high humidity and considerable rainfall, but will grow reasonably well without these conditions, and can even be expected to grow moderately well on chalk. It is fairly hardy and has no climatic limit to survival in Britain although, where foliage is scorched through cold or hot dry winds, performance will be impaired. Very young plants may be frost-tender. It is wind-firm and unlikely to snap. Deer may be attracted to red cedar's evergreen and scented foliage and cause damage. In the nursery, *Keithia* disease, caused by the fungus *Didymascella thujina*, is a serious pest. It thrives in *Thuja* hedges and leaf litter, and for this reason such hedges are never grown around tree nursery beds.

Propagation and management

Red cedar seed is very small indeed, but even so germination is generally good. Collections can be made once the trees are over 15 years old. Young seedlings have juvenile foliage at first which consists of single sharp needles standing out from the shoots. As an alternative to seed, cuttings can be easily rooted in July using bottom heat and mist. Planting out and establishment present no problems, as the young trees are very different in appearance to normal forest weeds, and are easily distinguished. They are shade-tolerant and can be underplanted, especially under birch or thinned-out larch. Although pruning is not essential, demand from florists for green foliage may make this operation, or an early thinning, worthwhile. Protection from rabbits and other vermin is essential. Plastic treeshelters have been tried and, although the trees usually survive in them, this method of protection is not really satisfactory for *Thuja*.

The nutrient requirements are low. The tree is widely naturalized in Britain and regeneration on wood edges and in clearings may be very thick, Unlike many other naturally regenerating conifers the western red cedar often survives the first few difficult years and successfully reaches pole stage.

▲
Red cedar forests occur across the whole country, including here, at Thetford Chase.

Timber quality and uses

Western red cedar wood is strong, light, and durable and is probably best known as 'cedar' roofing shingles. In addition, it is used for timber-clad houses, sheds, conservatories, and fencing. In milling, knots may be a problem and large stems are deeply fluted. The wood may also tend to collapse when kiln-dried. The fragrant foliage may be harvested for floristry. As a live tree it is an ideal ornamental specimen tree, and various cultivated forms have been developed. It was widely used for evergreen hedging prior to the introduction of

Leyland cypress. Its tolerance of a wide range of soils has led to its frequent use in small woods and shelterbelts around farms, where it survives in company with other species and provides year-round cover, and ultimately a crop of durable poles.

The timber is attractively coloured and totally durable out of doors.

8

Wild Cherry or Gean

Prunus avium

L.

Wild Cherry or Gean

Prunus avium L.

Prunus is the classical name for the plum which along with the peach, almond, cherry, and laurel make up this genus. The species name *avium* is from the Latin *avis* meaning 'bird', and birds are greatly attracted to the fruit. The common names are self-explanatory. Gean, which is a name used widely in Scotland, comes from an Italian name *Guina* for a local variety of cherry.

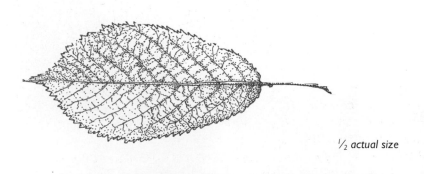

$^1/_2$ *actual size*

Origin and distribution

Prunus avium is Europe's tallest cherry tree often reaching heights of 30 metres (98 feet) on good sites. Its natural range includes southern Britain and most of Europe, but there is some doubt about the real geographical origin of the tree. Some authorities consider it to be wild only in west Asia from where it was probably brought into Europe by early man. The details of the introduction or reintroduction of the cherry to Europe may be debatable, but progression into the southern half of the British Isles seems to have been quite unaided by man some 8000 years ago, shortly after oak and ash. Scottish trees do appear to have been introduced ahead of any natural progression, but have become fully naturalized. Fossil remains in Britain of earlier cherry populations have been found, confirming the existence of a preglacial population here. The earliest British historical record for the tree is 1634. The wild cherry has been subjected to rigorous selection, mostly for the improvement of the fruit quality. The scientific species name and status have until comparatively recently been in some dispute; in 1886 George Bentham and Joseph Hooker, with reservations, placed *P. avium* as a variety of *Prunus cerasus*, the sour cherry. *P. cerasus* var. *avium* was described, quite mistakenly, as the variety without suckers. Alfred Rehder in 1940 failed to mention the species epithet *avium* at all. The original Linnaeus species designation for two distinct species *P. cerasus* and *P. avium* is now considered to be correct. The largest British trees on record are 31 metres (101 feet) tall (now felled) and 170 centimetres diameter at breast height (17 feet 6 inches in girth) in 1984.

Key features

In the wild, root suckers may often develop around established trees, which eventually grow as large as trees themselves. In this way, small woods may emanate from a single tree. Cultivated cherry trees, which are usually grafted on to wild cherry seedling rootstocks, also sucker freely if not restrained. Cherry bark has purplish-brown and pale pinkish-brown horizontal bands, which have a lustrous sheen on young trees. The tree carries well-defined lenticels in horizontal corky bands, and with great age trunks may gradually become rough and vertically fissured. Stem and branch wounds may exude a curious, clear, yellowish gum reminiscent of resin in conifers but without the familiar resinous smell. Buds are alternate, oval, and pointed, each having well-defined red-brown scales. The elliptical, sharply pointed leaves have distinct teeth and are carried on long glandular stalks. After a summer of dull mid-green, the leaves turn to orange and then, for a day or two in October, radiant crimson-purple. In

▲
The leaves turn to vivid reds and purple tints in autumn.

◄
The flowers hang in small clusters before the petals shower down like spring snow.

late April the nectar-rich flowers occur in clusters of three or four, each 2-centimetre flower having five white petals and many yellow prominent stamens. The long flower stalks arise from short spur shoots all over the crown of the tree, but never on current shoots. Each flower has a single seed chamber which develops into the familiar cherry stone, and the surrounding fruit is pulpy and yellow. As ripening progresses, the glossy skin goes from yellow, to red, and finally to black. The sweet cherries are attractive to blackbirds, which rapidly strip the trees of fruit. Stones that pass through birds are believed to germinate the following spring; otherwise they may lay dormant on the ground for nearly two years. Seedlings have two distinct, round, fleshy seed leaves which appear before the true leaves.

Growth conditions

Cherry trees can be found in almost every part of the British Isles at fairly low elevations. Best results in terms of timber production and quality are obtained from well-drained agricultural soils. Although the species is not really exacting with respect to soil, good, lime-rich, heavy, moist loams appear to be ideal. The tree requires full sunlight and is only shade-tolerant in its early stages of development. Very late frosts may spoil the flowers and young foliage, but damage is seldom serious. A wood which has originated as suckers from only one tree may be self-sterile and produce no seed.

Propagation and management

There are about 6600 clean cherry stones in every kilogram of seed. Although germination is normally about 80 per cent, this may take 2 years to achieve. Up to 1100 usable plants may eventually be expected after two or three years in a nursery. A tree will usually produce its

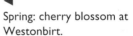

Spring: cherry blossom at
Westonbirt.

first good crop of cherries after about 25 years, as small early crops are usually taken by birds
before they can be picked. Collections from suitably protected trees can be made from August,
and heavy seed crops occur about every third year. Newly planted trees require protection
from animals and respond well to the total removal of weeds, either by chemical control or
mulching. Trees should not be allowed to shade each other out; lower branches should be
pruned off to give clean stems, as this will ultimately contribute to timber quality. Low-value
sapwood makes up a substantial proportion of small poles, so early thinnings are usually of
little value. Large diameter poles are likely to be worth the wait.

Timber quality and uses

In the past, most cherries were planted as a food crop, although the timber also has value. It works and polishes well and is a warm, reddish-brown colour tinged with streaks of gold and pale green. It has an intricate, attractive figuring and veneers carry the swirling patterns of grain. It has always been used for high-quality furniture and is still in demand for this purpose today, and is much prized by cabinet makers and musical instrument manufacturers. In former times it was used to make cask hoops and vine poles. Smokers used to cherish their cherrywood pipes and a few are still made. Cherry firewood burns well, wet or dry, and produces perfumed smoke. *P. avium* is one of the parents of the domestic black cherry, and is frequently used as a root stock for ornamental flowering and fruiting orchard varieties.

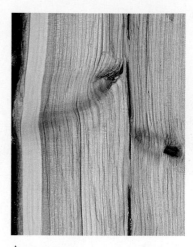

▲
The timber exhibits an extraordinary array of colours.

9

Sweet chestnut

Castanea sativa

MILL.

Sweet chestnut

Castanea sativa MILL.

The Romans called this *Castanea* from Castanum, a town in Thessaly where the chestnuts grew in abundance. *Kastanon* is the Greek word for chestnut. *Sativa* is from the Latin, meaning cultivated. *Chesten-nut* was the early English name for the tree and the fruit.

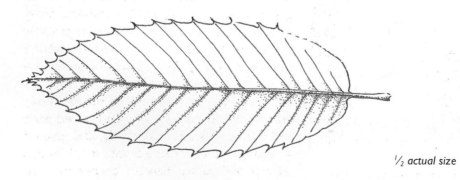

¹⁄₂ *actual size*

Origin and distribution

The early history of this tree was uncertain for a long time, but it has been established that it is not a native species but was introduced to Britain and Europe. The Romans are credited with the British introduction, presumably for the sake of its edible nuts, but the Greeks brought the tree to Italy from its native habitats in Asia Minor. As long ago as the third century BC, Pliny records that several varieties had been selected for their superior nuts. Chestnut charcoal has been found in British archaeological sites which date prior to the Roman period, but it is not known whether the tree was growing in Britain or had been imported as timber. By the reign of Henry II, the use of chestnut timber was widespread, and there is some evidence that it was even being exported to Normandy. The tree is now found all over Britain, but it is only naturalized in the south and east. The largest specimens are in Dorset (426 centimetres diameter, 43 feet 10 inches in girth), and the tallest is in central Scotland (36 metres tall, 118 feet).

▲
The tough leaves have saw-like teeth.

▲
Chestnut flowers are complicated structures, often female at the base and male towards the top.

Key features

Sweet chestnut is a large, tall tree with a domed crown, good green foliage and yellow autumn colour. It has deeply fissured bark which can be strongly vertical or somewhat spiral in either direction. The heavy, smooth branches support stout twigs which are rather angular, rough, and marked by distinct ridges. Fast-grown coppice shoots in particular have distinct cream-coloured lenticels. Later in the year a greyish bloom develops up the current stem. The winter buds are large with just a few yellowish-brown scales that have a rosy tinge on the upper side. Although pointed, they do not come to a sharp tip. The leaves are quite distinctive: tough, 20 centimetres (8 inches) long, elliptic, and a rich green. They have about 20 prominent, fairly parallel veins, each leading to a large softly spined tooth. These regular teeth resemble a sawblade. The pale yellow leaf stalks are usually 3 centimetres (1 inch) long and the leaf base may be cuneate or obtuse. Long creamy-white male catkins and mixed male and female catkins open in July, and give off a strong, rather unpleasant smell which attracts pollinating insects. Females are clustered at the base where one or a small group will develop into four-sided, sharply spined fruit or burrs up to 6 centimetres (2 inches) across. The fruit, which ripens in October, is the familiar glossy golden brown chestnut, with 1–4 nuts in each burr. Tree growth is very rapid in the early stages, particularly if grown as coppice, and trees have the potential to live for a very long time.

Growth conditions

Well-drained, moist, loamy, acid sites for chestnut trees can be found throughout Britain in sheltered valleys and lowlands. The species will not usually grow well on neutral or alkaline soils. Very large trees are seldom free of defect and often have little or no commercial value, so chestnut woods are mostly retained for amenity shelter or sporting use. Chestnut coppice, on the other hand, is worthwhile, but ideal coppice sites are

▲
The familiar chestnut fruits are guarded by viciously spined husks.

Chestnut woodland, as here in the Forest of Dean, is often overgrown coppice.
▼

confined to a small area of the country—parts of the Kent and Sussex Weald are best. Woods of fast-growing stored coppice stems which are thinned out and allowed to grow on to tree size, are often found over a much wider area. The nuts reach edible size only in the south, although proud owners further north regularly boast of good nuts from their trees. This species requires a higher degree of acidity in the soil than most other broadleaves will tolerate, and thrives on clay with flints and sandy sites. Once established, its heavy shade and tough leaf litter will kill most ground vegetation in the immediate vicinity. Although coppice will tolerate a little shade, its vigour is reduced by this and the very best growth will be found only in full light.

Propagation and management

The sweet chestnut is easily raised from seed, but storage of nuts over winter must be carefully done, as drying out, vermin, or mildew may cause serious problems. Planting presents no particular difficulties although weeds, of course, must be controlled, and rabbits, hares, and deer not allowed to damage the young plants. Voles may be a problem on grassy sites. The chestnut suffers from few diseases and is hardly ever damaged by hot, dry weather, although the fungus *Phytophthora* has been a problem on wet sites. The tree is widely grown for shade or ornament on sites that would not normally be expected to support broadleaved trees at all. In some circumstances, it may be considered an acceptable visual replacement for elm.

Timber quality and uses

Chestnut timber is straight-grained and closely resembles oak in young trees. With age, however, it develops spiral grain and splits badly. Ring and star shakes develop in old trees; at the sawmill whole logs may simply fall apart. The wood has to be harvested young to be of value, and timber rotations of not more than 70 years, or coppice, are recommended. Quarter-sawn logs, although lacking the large medullary rays of oak, do have a prominent figure provided by the distinct brown annual rings. Chestnut timber is 20 per cent lighter in weight than oak, and if straight is easier to work. Drying is difficult and twisting or collapse may occur. The sapwood is narrow, so there is a large percentage of durable heartwood; it is good in contact with the ground but corrodes iron nails and fittings in wet conditions. The wood peels well to make plain veneers. There have been numerous traditional uses for small chestnut poles. When used for cask staves the wood improves the colour, sweetness, and longevity of wine. A chestnut charcoal industry, in its day, provided coals for hop kilns and iron smelting. Chestnut bark has been used for tanning but could not compete with oak for quality or ease of harvesting. As firewood it gives good heat but spits and sends sparks flying

▲
Good-quality timber is highly prized and closely resembles oak.

from an open hearth. The main traditional product of coppice was hop poles when the demand was great, but high-tensile wire and fewer home-grown hops have led to a decline in this market. Since 1905, cleft chestnut paling fences have been a major coppice product.

▲
The chestnut has a long tradition of being grown as coppice to produce small poles.

10

Lawson's cypress

Chamaecyparis lawsoniana
(MURRAY) PARLATORE

Lawson's cypress

Chamaecyparis lawsoniana (MURRAY) PARLATORE

The genus name is from *chamae* meaning 'ground-hugging', the Greek *kuo* 'to produce', and *parisos* meaning 'equal' or 'symmetrical' for its similarity to the true cypress *Cupressus*. Lawson's and *lawsoniana* commemorate the Edinburgh nurseryman Peter Lawson, to whom seed was first sent in 1854.

⅔ actual size

◄
Most Lawson's cypress trees are hidden deep in forests, so their size may not be readily appreciated.

Origin and distribution

The Lawson's cypress tree is native to two restricted highland regions of south-west Oregon and north-west California. It was introduced to Britain from the upper Sacramento Valley in California in 1852 by John Jeffrey, and then in 1854 and 1855 by William Murray. In the wild it will grow to 50 metres (164 feet). In the British Isles the tallest specimen was 40 metres in 1987 (131 feet) at Balmacaan in the Scottish highlands. The largest breast height diameter known in 1989 was 158 centimetres (16 feet 3 inches in girth) at Powerscourt in County Wicklow. It is widely grown in parks and gardens and there are a few small areas of forest throughout the country.

Key features

The tree is upright and columnar, but does not always confine itself to a single stem. The mature bark is fairly soft, rather stringy, and may peel off in long vertical strips. On young trees the bark is smooth, greenish-brown and somewhat shiny, becoming purplish or reddish-brown with age. Old trees are ridged and sometimes heavily fluted towards the base but usually less so than the western red cedar. The shoot is dullish green, almost completely obscured for several years by densely packed tiny scale-like leaves in opposite pairs. Each leaf or scale has a translucent gland in its centre which coincides on opposite leaves to give the impression, when viewed through a lens against the light, of piercing the central shoot itself. The pointed scale tips are incurved and so small that the

▲
Lawson's cypress probably has a higher number of leaves and fruits per tree than any other in Britain.

foliage does not have a spiky appearance. Crushed foliage has a bitter smell, and some people are allergic to the oil it produces. Dry foliage burns fiercely even when it is still green. Numerous, tiny, globular male flowers appear on the tips of weak shoots in late winter, starting off almost black and turning brick-red in April as they become ripe. Clusters of females develop into hard, round, pea-sized cones which are green with a dusting of white at first and then become purplish-brown. Each cone contains about 10 minute seeds, each of which has a pair of tiny lateral wings.

Growth conditions

Experiments in Wales and south-west England indicate that this species has almost the lowest nutrient requirements of any tree planted in upland Britain. It will grow and thrive even where the Sitka spruce comes under stress. It is resistant to atmospheric pollution, but may suffer foliage browning in severe winters or excessively hot summer sunshine. Low temperatures in Britain do not kill it, and even frost damage is usually only temporary. Although it is unsuitable for poor peatlands and dry heaths, it will grow almost anywhere else. It tolerates some shade and responds well if underplanted, but growth will be slow where shade is heavy or where overhead cover begins to touch it. Trees rarely blow down, although strong winds or heavy snowfall will prize apart weak forks as the stems gain weight of foliage. Ice and snow may bend young saplings right down to the ground, but they invariably recover completely as soon as the temperature rises. The dense evergreen foliage provides good cover and nesting sites for small birds and is a home for numerous insects. The forest floor, although laid bare of live vegetation, becomes covered by a litter layer of small dead sprays of foliage which provide

further protection for small animals and insect life. The tree is an effective windbreak and provides good shelter on its leeward side.

Propagation and management

Seed production in Britain is abundant and germination is outstanding. Due to hybrid variability, some home-produced seed may produce unexpected variants. If identical individuals are required on a small scale, cuttings should be used for propagation, which is achieved under glass in summer using mist and bottom heat. Young plants grow rapidly and should reach plantable size in only two years. The species is not widely used as a forest tree in Britain, perhaps because western red cedar is more productive and less likely to produce multiple stems. Lawson's cypress can be grown to provide foliage for the floristry industry. The plant is green throughout the winter, so protection may be required from deer and rabbits. There are no serious insect or pathological problems in Britain.

Timber quality and uses

Port Orford cypress is both the tree name and the commercial lumber name in America, where it is an important timber. The wood is yellow and brown coloured, close-grained, and strong. It has a clean 'pine' smell and may be resinous. In Britain it is mainly used for small, light poles, although it was once prized for ladder poles. As rustic garden fencing cypress wood is ideal, although not entirely resistant to decay. As a hedging plant it is also useful where rapid growth in the early stages is not a priority. A dense, head-high hedge can be expected in ten years which will tolerate regular clipping. Beyond this size, Lawson's cypress hedges are particularly valuable because their annual growth rate is slower than other similar types of conifer hedging. There are numerous ornamental forms that are outstanding garden specimen plants. They come in every texture, size and shape, and in several colours from blue to gold.

▲
This is an important commercial timber in America and a high quality can also be achieved in Britain.

11

Leyland cypress

× *Cupressocyparis leylandii*

(DALIM. AND JACKS.) DALLIMORE

Leyland cypress

× *Cupressocyparis leylandii* (DALIM. AND JACKS.) DALLIMORE

Cupressocyparis is a modern amalgamation of the parent genus names. *Cupressus* from the Greek *kuo* meaning 'to produce', *parisos* meaning 'equal'-(sided tree), and *chamaecyparis* meaning a 'small cypress'. *Leylandii* commemorates C. J. Leyland, owner of Haggerston Hall when the first clones were grown there. His brother-in-law was Mr Naylor of Leighton Hall. 'Naylor's Blue' and 'Leighton Green' are subsequent cultivar names.

¹/₂ actual size

Origin and distribution

The Leyland cypress is a spontaneous bigeneric hybrid between Nootka cypress and Monterey cypress (*Chamaecyparis nootkatensis* × *Cupressus macrocarpa*). Although the parent trees are both native to North America, the cross has never been reported from there, and is only known to have occurred unaided in the British Isles. The resulting hybrids are sterile and can only be propagated vegetatively. Six clones originated from seed collected from a Nootka cypress at Leighton Hall in Powys in 1888 and were planted at Haggerston Hall in Northumberland. In 1911, two more clones came from a reverse cross at Leighton, but these were not recognized until 1925. A further two clones, from Monterey cypress seed collected in 1940 at Stapehill in Dorset, were subsequently isolated, but the parentage of these is uncertain and may involve Lawson's cypress. Forest plots of Leyland cypress are few at present, but the potential timber production from this hybrid is enormous, and more plantations could be worthwhile. Among the green clones of the tree, 'Leighton Green' and 'Haggerston Grey' are most commonly planted. The former has reached 33 metres (108 feet) by 127 centimetres diameter (13 feet 2 inches girth at breast height). In 1990 a 'Haggerston Grey' was measured at a massive 36 metres (118 feet) by 97 centimetres diameter (10 feet in girth) at Bicton in Devon.

This plantation is only 35 years old.
▼

▲
Leyland cypress foliage is difficult to distinguish from other cypresses, but the cones are larger.

▲
Under a mass of evergreen foliage, large rugged trunks are concealed.

Key features

Huge specimens are commonly seen, showing considerable vigour, and with dark green, columnar crowns which usually obscure the whole trunk from ground level. The general outline of the tree is spiked as each shoot tip extends rapidly upwards. The bark is smooth, purplish- or greenish-brown, becoming shallowly ridged and a little stringy with age. If cut or damaged it is strongly resinous and has a bitter smell. The foliage consists of long heavy sprays holding countless tiny green scales for many years. Bare branches and dead wood are seldom seen, occurring only in the crowns of very old trees and then usually obscured by healthy exterior foliage. The cones are dark, globular, and shiny brown with eight bluntly spiked scales. Although there are several seeds between each scale, these are usually sterile. If by chance any seeds should germinate, the progeny are unlikely to resemble the parent tree.

Growth conditions

The Leyland cypress hybrid has all the virtues of its parents but is stronger, faster-growing, and hardy. Most clones grow very rapidly indeed and are resistant to air pollution, salt spray, drought, and cold winters except in very windy or elevated situations. Widespread garden use has exposed the tree to every soil type, fertilizer, and pesticide drift without causing much harm. It is lime-tolerant in lowland gardens and acid-tolerant on mountains or sandy sites.

▲
A clonal forest consists of identical individuals.

Despite the perpetual gloom at lower crown level and under the trees, unthinned plantations appear to contain surprisingly few suppressed or dead stems. *Leylandii* seldom blow down, except on very wet or soft ground; however, with such a dense, extensive, wind-resisting crown and strong, spreading root system, some stem snap is inevitable. This may occur at any weak point on the tree, from the stump upwards. In environmental terms, plantations have little to offer except evergreen shelter to animals and birds. No plants can coexist with Leyland cypress once the canopy has been closed.

Propagation and management

Little is known about this hybrid in plantation conditions, but the few small plots that do exist appear to present no serious tree health or production difficulties. Clonal stock is very regular in speed of growth and visual appearance, and thinning out is usually possible on a geometric basis. Plants have to be raised from cuttings; best results, outside of micropropagation, are achieved when cuttings are inserted into heated greenhouse beds in February, March, or September. Semi-ripe wood about 15 centimetres (6 inches) long taken from upward-growing shoots on stock plants is best. Cuttings should be dipped into a root-promoting growth substance and inserted vertically close together, and watered at first by overhead mist or fogging. After one season plants will need to be potted on or bedded out. In strong winds young plants of some clones may be damaged by rubbing shoots together or against adjacent plants or weeds. In southern England Stapehill 'Clone 20' is vulnerable to hot dry summers.

▲
The timber is strong and durable, and very quickly produced.

Timber quality and uses

There is no wood processing industry for Leyland cypress although there may be potential for one in the future. The timber is heavy and resinous. Colours include pale brown, creamy-white, and yellow. Foliage is equally important as Lawson's cypress for the floristry market. The real value of the tree at present is as a living specimen, a shelterbelt, hedge, or landscape feature. As a hedge it can be clipped annually with little browning or loss of vigour. As a specimen tree it can reach a large size in a short time (ten metres in ten years). Golden foliage forms have been developed using the golden Monterey cypress as one parent. Gold forms also grow quickly 'Robinson's Gold' and 'Castlewellan' have both exceeded 11 metres (36 feet) in height and annual shoots in excess of 60 centimetres may be expected.

12

Douglas fir

Pseudotsuga menziesii
<small>(MIRB.)</small> FRANCO

Douglas fir

Pseudotsuga menziesii (MIRB.) FRANCO

Pseudotsuga is taken from *pseudes* meaning 'false' and the genus name *Tsuga* which Douglas fir resembles. It was discovered by, and named after, Archibald Menzies. The common name commemorates David Douglas who introduced it to Europe. It was initially assigned to the firs (*Abies*).

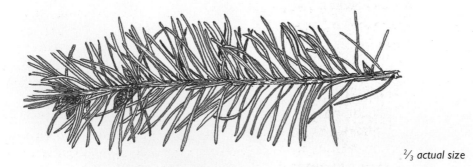

$^2/_3$ *actual size*

Origin and distribution

The natural distribution of the Douglas fir extends from British Columbia to California, from the west of the Rocky Mountains to the Pacific Ocean. The variety *glauca* extends further south into Mexico. It was introduced to Britain by David Douglas late in 1827 and several original trees are still standing and growing strongly. It has been widely planted throughout the country, where it is usually reserved for good quality forest land and lower hill slopes. It has always been a favourite ornamental park and arboretum tree; consequently the best and largest specimens are still to be found in old estate woodlands. The tallest trees are found in Tayside, where in 1985 some had reached 61 metres (over 200 feet). The largest trunk diameter at breast height by then was 215 cm (22 feet 2 inches in girth). The optimum size and age in Britain is unknown and has yet to be reached.

Key features

This impressive, straight-stemmed conifer forms lofty forests in western Britain. As a specimen tree it usually becomes a huge conical pillar with hanging curtains of rich green foliage, often reaching to the ground. Storms take their toll on the trees as they age, giving them a gaunt appearance. The bark on old specimens becomes rough and corky, cut vertically by haphazardly broken fissures. The overall stem colour may vary from tree to tree and includes shades of light and dark purplish-brown, grey, and even orange. Young trees and upper branches have smoother, dark, greenish grey-brown bark with numerous penny-sized resin blisters. When burst by an axe or saw, these produce a

▲
The needles are soft and smell of the forest.

shower of resin which smells strongly of turpentine. The young shoots are yellowish-green with short, fine hairs. The 2.5 centimetre (1 inch)-long soft needles are solitary with a blunt, but unnotched tip. When a needle falls (usually after three to five years) the leaf stalk, or pulvinus, goes with it, leaving a fairly smooth scar on the surface of the shoot. The deep green or grey-green foliage has a strong, sweet, fruity aroma, especially on hot days. The buds, which are reminiscent of beech and hornbeam, are slender, sharply pointed, ovoid–conic, pale, glossy brown to bright red-brown in colour with unfringed scales. The flowers look similar to those of other conifers: the males consist of clusters of stamens maturing in March and April; the females, growing nearer the branch tips, are small, untidy, pinkish-green cones. The cones themselves mature in one season, and become pendulous, dark brown, and about 8 centimetres (3 inches) long, with distinctive three-pronged bracts projecting between each scale. The winged seed is shed in September and October of the same year.

◀ Male and female (right) flowers occur together in spring.

◀ The cone scales have distinctive trident bracts projecting from them.

Growth conditions

Although the Douglas fir is a fairly tolerant species, it is neither as hardy as the Scots pine nor quite as fast-growing as the Sitka spruce. It will not always stand firm on wet clays and is likely to be damaged by severe exposure or late frost. It prefers mineral soils to deep peat, and is very much at home on rocky hill sites. It grows well on good, fertile ground, but may be deformed or killed in early life by competing vegetation such as bracken or honeysuckle. Young trees are shade-tolerant and grow well in an underwood situation. On maturity the tree tops in a forest become uneven, and dominant specimens protrude above the rest which may ultimately be blasted by strong winds and become 'stag-headed'. Thicket-stage Douglas fir plantations on lowland sites can coexist with a rich and diverse ground flora (except for honeysuckle) and for ten years provide good cover for small nesting birds and game. Deer are

▲
Fine timber trees of Douglas fir grow in much of western Britain.

attracted by the smell of the foliage and may do considerable damage to young trees. Later on, as the forest becomes dense, almost everything except ivy is gradually shaded out. However, woodland birds including tree creepers and nuthatches value the rich population of insects in the foliage and on the rough bark. On suitable sites natural regeneration is likely to appear around the fringes of the wood and into any windblown gaps that occur.

Propagation and management

Seedlings of Douglas fir grow rapidly in an open nursery and should reach a plantable size in only one year. One kilogram of seed, containing some 70 000 individual seeds, may be expected to produce 24 000 usable seedlings, providing that the tiny gall wasp *Megastigmus spermotrophus* has not infested them. Seed production begins after about 30 years and peaks at 50 years, and cones must be picked in September before they open. Grafted 'orchard' trees simplify cone collection, ensure high quality, and reduce the time required to produce seed. Orchard cones can be picked from ground level, whereas 50-year-old trees bear cones much higher up. On less than optimum sites even established tree growth can be severely reduced by the Douglas fir woolly aphid, *Adelges cooleyi*. At every age, plantations are subject to the danger of forest fires as the foliage burns fiercely, especially during periods of hot dry weather. Wood production is rapid in healthy young trees but tends to slow down after about 50 years. Where growth is particularly good, thinning may begin 20 years after planting, giving an early financial return on small poles. Instability is common on wet ground, especially after a delayed thinning.

Timber quality and uses

Imported Douglas fir timber has the rather confusing trade name in Britain of 'Oregon pine'. However, home-grown material is now usually simply referred to as 'Douglas fir'. It has a well-defined, undulating, reddish-brown heartwood and conspicuous brown annual rings which show up very clearly and prominently on flat-sawn surfaces. The sapwood, by contrast, is a pale, creamy-buff colour. The wood works well, is very strong, and a little heavier than Scots pine. Large, slow-grown, knot-free trees produce straight timber of very high quality. The Douglas fir has numerous uses including large-size building material, particularly for internal structural work that will remain exposed to view. Although it stains and takes paint

well, treatment with preservative chemicals is difficult, because the hard annual rings resist penetration of most substances. Functional, as opposed to decorative, plywood is made from Douglas fir veneers, as are laminated beams. There are numerous traditional estate uses of Douglas fir timber, particularly for small poles. It makes good pulp for the packaging industry, and in former times it was commonly used for telegraph poles and railway sleepers. Large bulks of Douglas fir timber were once extensively used for dock and harbour work, and for piling.

▲
The wood is brightly coloured and heavy.

13
Grand Fir

Abies grandis
(DOUGLAS) LINDLEY

Grand fir

Abies grandis (DOUGLAS) LINDLEY

Abies (from Latin *abire*, 'to go away') refers to the speed at which this tree grows. *Grandis* means large and lofty. The common name comes from the same origin.

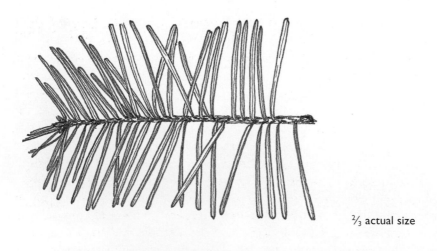

⅔ actual size

Origin and distribution

The grand fir is the tallest of the true firs, reaching 100 metres (328 feet) in its native range in north-west America. Its natural habitat extends from the northern end of Vancouver Island south to the Navarro river in California, and from the Pacific coast to Idaho and Western Montana. David Douglas introduced it to the British Isles in 1831, six years after he discovered it on the Columbia River. The last of the original trees to survive are at Curraghmore in County Waterford and at Lochanhead, in Dumfries. A later introduction by John Jeffrey in 1852 has many more survivors. This species has been widely planted throughout the British Isles, but grows best in the cool, wet climate of the west. Growth is exceptional and the grand fir will outclass almost every other conifer on any forest site in western Britain. Heights of 30 metres (98 feet) can be expected in about 35 years. There are plots and small plantations in most forest areas and specimen trees occur in many tree collections. The largest known specimens were 52 metres (171 feet) tall in 1990 and 223 centimetres in diameter (23 feet in girth).

▲
This stem is 115 years old, found at Westonbirt Arboretum in Gloucestershire.

Key features

Where it is suited to its environment this tree is narrow and columnar, with a spire-like crown of drooping branches with upturned tips. The foliage is deep green, but when the wind lifts the branches the pale undersides of the needles shine out. Green foliage tends to extend some way down the stem because of the shade tolerance of the tree. The trunk is persistently straight and the horizontal branches are generally thin. The bark on trees less than 30 years old is smooth, dark greenish-brown, and pebbled with resin blisters. Older bark becomes thickened and corky with broad, deep grooves between prominent plates of silver grey. The shoots are olive green with a few hairs and shallow longitudinal grooves. The aromatic foliage is widely spreading in two horizontal ranks along the twig. The single, flattened, linear needles vary in length but commonly reach 3 centimetres (1 inch). The tips are rounded and notched, and the leaf stalks are sharply twisted. The needles are shiny and dark above and the undersides have two prominent bands of greenish-white stomata. The buds are small, globular, and resinous; the scale tips are sometimes free. Male flowers are very small and purplish until they become yellow with pollen in May. They are densely packed under short sideshoots. Females are confined to the tree-tops. The deciduous, erect cones may be 10 centimetres long (4 inches); they are light green at first, and then become dull brown with frequent blobs of resin. Once the cone disintegrates, some four weeks after ripening, the central vertical core, a tough spike, remains on the tree. The seed is triangular with a firmly attached wing.

▲

The needles are in two ranks along the shoots.

Growth conditions

For best results in Britain, this tree should be planted in areas of high rainfall. Even so, trees planted in less than perfect conditions frequently do well. In the north and west the grand fir will thrive on good Sitka spruce sites. A large proportion of the green foliage is retained for several years. In the gloom beneath plantations there will thus often be no ground vegetation at all. A typical stand of grand fir will contain trees of various sizes—sub-dominant trees coexist with the strongest specimens. Rooting is generally very good and, unless impeded by heavy stocking or lack of drainage, will hold the trees firm in strong winds. Stems frequently snap, however, because the trunks can be very slender and the wood is soft and weak. Stem crack is sometimes worrying for the forester and timber merchant but it is less severe in this species than the noble fir. Needles retained for five or six years may become very sooty near industrial areas. The species is completely hardy in Britain, but foliage may be scorched by dry or very cold winds.

Propagation and management

Seed can be collected from the tree only in the very short period between the ripening and disintegration of the cones. The best imported seed comes from Vancouver and Washington State. Viability is variable and estimates of plant production are usually unrealistic. Sowing should be earlier than for most other conifers. Growth in the open nursery is often slow, and it may take four years to obtain a usable plant. Young plants will tolerate a great deal of shade and take very well to underplanting. When planted in trials, under widely spaced mature larch trees, grand fir tends to grow rather faster than on adjacent open ground. Competition from heather, *Calluna vulgaris*, is far less of a problem than it is for the Sitka spruce. Protection must be thorough: deer in particular are attracted to the scented evergreen foliage and do considerable damage. There are no serious diseases specific to this species. It will not thrive on highly calcareous soils or poor dry heathlands.

Timber quality and uses

A huge quantity of timber is produced by grand fir in a very short time. The wood is soft and is not durable, but it is quite suitable for making boxes and pallets and for rough joinery. It makes good pulp, but the quantities available in Britain are not enough to sustain an industry in isolation. The wood is heavy when wet but very light when thoroughly dry. It can be treated with preservatives quite easily. The grain is straight but in long lengths may show a spiral tendency. Knots are generally small and usually tight, and stem taper is very limited. Some logs have cracks, which may be spiral, and are damaging to the wood. As an isolated specimen tree, this species is quite impressive with its great height and sweeping green foliage, which often reaches down to the ground. Unfortunately, in dry areas the needles can be blasted from the tops of the trees by the wind, making the trees 'stag headed' and then unsightly.

▲
The wood is soft and clean, and produced very quickly on a good site.

It may seem odd that the European silver fir, *Abies alba*, is not a common forest tree in Britain. Occasionally, very old giant individuals and even small plantations may be found. This is due to severe damage caused by the silver fir wooly aphid, *Adelges nordmannianae*. Repeated attacks cause the tree to die out completely.

▲
The Noble fir is a very hardy tree, here growing high in the mountains of North Wales.

14
Noble fir

Abies procera

REHDER

Noble fir

Abies procera REHDER

Abies comes from the Latin *abire*, 'to go away', referring to the rapid growth and height achieved by the firs; *procera* means tall. The common name is self-explanatory.

½ actual size

This stem is about 130 years old and one metre in diameter.

Origin and distribution

This handsome tree is a native of the high country of Oregon, Washington State, and California. It was discovered by David Douglas in 1825 and introduced by him to Britain in 1831 after a second visit to one of the noble fir forests. Subsequently in the 1850s, large consignments of seed were imported by the Oregon Association and William Lobb. In Britain, particularly the north and west, no tree has survived better than the noble fir. It will equal any conifer except possibly lodgepole and mountain pines on the most severe elevated sites. In wet areas of Britain, it will thrive in deep peats and at higher elevations than almost any other tree species. Unfortunately, it may be subject to attack by aphids, which rules out its use in some areas. Several individual trees have reached 50 metres (164 feet) in height in the Strathclyde Region, and the largest stem diameter known in 1987 was 187 centimetres (19 feet 3 inches in girth). One per cent of conifer forests in Britain consisted of this species in 1990.

Key features

As a specimen tree, the noble fir is only slightly less impressive than the grand fir. The blue form 'Glauca' is particularly decorative. In even-aged fir woodland, an obvious and important feature is the enormous range of tree sizes that coexist. Another feature of the noble fir is the preponderance of dark, horizontal branches that appear to be evenly spaced up the stems—internodal branching hardly ever occurs. The bark is silvery-grey to dull, purplish-grey on some younger stems, and is smooth at first with prominent resin blisters. The bark will become fissured later on but always retains some smooth silvery bark between the cracks. The crown is conic, and green foliage is retained to the ground on open-grown specimens; even in plantations foliage extends well down the tree. In time tree tops may be lost through heavy coning and strong winds, either flattening out or becoming 'stag headed'. The leathery needles are flat, upswept, and curving in two directions to form more or less horizontal ranks. The twig is reddish-brown and covered in fine hairs, but is almost entirely obscured by tightly packed twisting needle bases. The needles are grey-green above and bluish-green below, giving the tree an overall dusty-green appearance. The flowers are similar to those of the grand fir, but the pale orange-brown cones are much larger and very distinctive: they stand on the top branches of the tree like huge, fat candles up to 20 centimetres (8 inches) long. In one season (May to October) they ripen and disintegrate, leaving only a slender, tapered, woody central spindle on the tree. Each cone scale has a reflexed bract extending well beyond it and covering the scale below. The whole effect is feathery, but in an orderly diagonal pattern.

▲
Each needle of the noble fir is smoothly curved to make it face up to the light.

Growth conditions

The noble fir is a mountain species that has low heat requirements and low evaporation rates. It is damaged by late spring frosts only when planted in deep hollows. Its nutrient requirements are modest and it will thrive on peatland, particularly flushed amorphous peats. It needs a lot of moisture but will survive on less than the Sitka spruce requires. It is shade-tolerant but less so than the grand fir. It is moderately wind-firm and usually resists snow damage until it is very old. It is completely naturalized in many upland regions and regenerates freely by seed. It is susceptible to industrial pollution and intolerant of lime-rich soils.

▲
In spring, the seldom-seen flowers of the noble fir are quite spectacular.

Propagation and management

Noble fir trees produce good seed early in life, but the cones are confined to the tree tops and must be picked by the end of August before they begin to fall apart. Original seed lots were imported from the Cascade mountains in Washington, and some still are today. Plants are slow to grow in the nursery, requiring two or three years to reach a size suitable for moving out into the forest. It is worthwhile considering underplanting as a possibility for this species; for example, young, thinned-out birchwoods provide ideal light, overhead cover.

▲

Noble fir cones are heavy and large (20 cm x 8 cm). They ripen in one season and then break up on the tree.

◀

The cones are only found on the topmost branches, often denuding them of foliage and weighing them down.

◀

The clean, white timber grows rapidly but has a tendency to crack.

Transplanting often sets bare-rooted plants back, and it may take two years or so for them to establish themselves out on the hill. Thinning young stands may be difficult because of the enormous range of pole sizes that usually develop together. Working in the forest with this species using hand tools is particularly unpleasant and dirty because of the sticky resin that showers from the cut bark and branches.

Timber quality and uses

The noble fir is a high-yielding species but the stem tapers markedly towards the top of the tree. The wood is soft, white, and not durable; it takes preservatives well. Air seasoning is quick and fairly easy, and dry wood is moderately stable. It works well, but spiral grain, if present, will tend to lift along sawn planks. Drought or frost cracks, which are also spiral, may be a serious problem for the sawmiller. Noble fir timber makes good, long fibre pulp, but in Britain the small quantity of available material is a limiting factor for the industry. The timber has no common name other than the tree name in Britain. As a live tree this species is preferred in Denmark and some other European countries as a Christmas tree. It has obvious amenity value but requires cool, damp sites and clean air in which to grow.

15
Hazel
Corylus avellana

L.

Hazel

Corylus avellana L.

The name comes from the Greek *korys* meaning 'helmet', a reference to the calyx covering the nut. The Greek word for hazel nut bushes is *karyon*. *Avellana* commemorates the town of Avella in Italy where the nuts were cultivated. Hazel is from the Old English, *haesel*.

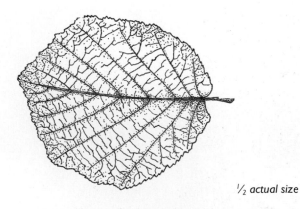

¹/₂ actual size

Origin and distribution

The hazel is a deciduous small tree, usually with multiple stems and a spreading habit. It is native to Britain, Europe, west Asia, and North Africa. In the British Isles it is common in all but the extreme north and west, and on particularly high ground. Most lowland soils are good for hazel growth. Chalk and some thin upland soils are tolerated, but the tree will usually be shrubby and unproductive in these conditions. The best ground for strong growth is where the soil is heavy but well drained. The hazel is a moderately shade-tolerant tree. It is not affected by even the harshest British weather, nor is it prone to any serious pests or diseases. It seldom grows taller than 7.6 metres (25 feet) and individual stems over 20 centimetres (8 inches) thick are usually at the end of their useful life.

Key features

The bark is smooth, except on very old stems, pale pinkish or grey-brown with scattered, dark, horizontal lenticels. Epiphytes usually obscure much of the underlying colour with patches of pale, jade-green, and dark browns. On old stems, small rolls of paper-thin bark peel back from vertical fissures but seldom fall off. Young shoots are pubescent, glandular, and pale brown. The leaves are roundish but abruptly pointed at the tip; the leaf bases are cordate and the petioles are short. Leaf margins are doubly toothed and bristly like the whole underside of the leaf, reminiscent of nettles or elm. The male catkins overwinter in a closed state and then open dramatically as the first warm days of spring arrive. Bunches of bright yellow, dropping 'lambs' tails' completely cover some plants. Female flowers on the same branches appear as tiny pink extensions to what seem to be slightly plump buds. The nuts are globose to ovoid, white at first and then turning to golden brown in September, still half enclosed in the ragged calyx which also turns from green to brown. Grey squirrels take most of the nuts before they have a chance to ripen, so natural regeneration does not occur where squirrels are common.

◄
Hazel nuts are usually taken by squirrels these days.

Hazel catkins ('lambs' tails') herald the spring.

Growth conditions

The hazel is a fundamental part of the British wildwood environment. Its ability to tolerate shade enables it to survive under the canopy of large trees, notably oak. As a tree it appears to be relatively short-lived, usually reaching about 60 years and then dying back or being prised out of the ground by the weight of snow on its wide-spreading branches. As managed coppice it can survive ten times longer and still produce a good crop of poles. The hazel can stand wind or frost but is frequently damaged by rabbits or deer in the first year after it has been coppiced.

Propagation and management

The management of hazel coppice and 'coppice with standards' is a precise and ancient science. The period between cuts, both of underwood and standards, is calculated to give the right size and type of produce from each. In addition, the timber

Traditionally, hazel is cut down to the base every 7–15 years. This clump is over 20 years old.

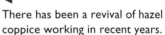
There has been a revival of hazel coppice working in recent years.

production goes on year after year without an unproductive break. Traditionally coppice would be cut on a 7–15 year rotation, and the area of woodland was divided into the same number of sections to be cut, thus enabling one part of the woodland to be harvested every year for all time. To avoid damage to the stools (rootstocks) and to obtain strong rods, cutting was confined to the winter months and would be completed by Lady Day (25 March). A proportion of overwood trees were cut to coincide with the coppicing. Conservation groups managing hazel coppice today tend to stick to these same principles. It was thought that chainsaw vibration would damage stools which had only been cut with a billhook for centuries, but fortunately this has been found not to be so. Efforts were made to manage hazel coppice economically in the 1950s without much success; from then on, hazel was grubbed out and poisoned on a large scale to make way for agriculture and plantation forestry. Today, surviving hazel woods are being saved by woodland trusts and conservation bodies. The hazel has, in fact, proved very difficult to kill, and considerable numbers of trees are reappearing in and around plantations of other species.

Timber quality and uses

Hazel timber, always produced in small diameters, is straight-grained, hard, and heavy, pale brown in colour with a delicate radial figure. The bark is thin and the heartwood is indistinct. There are no modern uses for it beyond craft items and fancy walking sticks. Hazel wood is not durable and must be treated with varnish or preservative. In the past, the hazel would provide most of the requirements of the rural economy, from edible nuts to firewood. It excelled as a hand-worked rod, either split or twisted. Thatching liggers, spars, sways, and pegs were made, as were hurdles for animal pens and garden fences. When finely cut it was used to make traps, cages, and cribs for animals, birds, and fish; long thin shoots provided 'heathers' for the tops of laid hedges. Hazel dowsing rods helped to locate water underground, and the charcoal was used to make fine artists' crayons and gunpowder. Hazel wood was ideal for the wattle of 'wattle and daub' which provided the base for the plaster between the timbers of lowly houses. In more recent times hazel wood has been used for paper pulp and fibreboard, but it is best mixed with softwood pulp for satisfactory results. There are still edible nut forms and a cultivar with ornamental purple foliage. The curious form known as 'Harry Lauder's walking stick' is the cultivar 'Contorta' discovered at Frocester in Gloucestershire in 1836.

▲
The timber as such is seldom used, but it is strong and straight.

16

Western hemlock

Tsuga heterophylla
(RAFINESQUE) SARGENT

Western hemlock

Tsuga heterophylla (RAFINESQUE) SARGENT

The name *Tsuga* is taken from the Japanese name for the Asian species. *Heterophylla* refers to the leaves being of different size on the same tree. 'Western' distinguishes the tree from the eastern form (*Tsuga canadensis*) in America. The crushed foliage is said to smell of hemlock.

⅔ actual size

◄
Mature trees develop spreading
buttresses and fluted stems.

Origin and distribution

This minor British forest species comes originally from the Pacific coast of North America, from Alaska to California and inland as far as the Rocky Mountains, where the tree has an elevation limit of about 1500 metres (5000 feet). It was discovered in 1826 by David Douglas and introduced to Europe in 1851 by John Jeffrey, and the first plantation was established at Inverary in 1888. In its native habitat on the best forest sites, the western hemlock may reach a height of 70 metres (230 feet). Before the last Ice Age, the *Tsuga* was extant in Europe, including Britain. Only the western species, however, grows straight enough to produce quality timber. The tallest tree in Britain is found at Benmore in Strathclyde which in 1991 had reached 52 metres (170 feet). The largest stem diameters in 1988 were in the region of 200 centimetres (21 feet in girth) in Tayside.

▶

The tiny, early summer flowers are a distinctive pale mauve colour.

▶

The leathery cones are small and freely produced all over the tree.

Key features

The western hemlock is straight even if sometimes multi-stemmed, with light horizontal branches which droop at the tips. Even the leading shoot is floppy during its first summer. On favourable sites, the rate of growth is enormous (70–100 centimetres per year for 20 years). However, because of the habit of internodal branching, this is not as easy to appreciate as the growth of spruces or pines, which produce more or less a single whorl of branches each year. Throughout the tree's life, the crown is a tidy cone of dense, dark green, down-turned foliage. New shoots are pale brown, ribbed, but almost hidden by rather flattened, short, soft needles arranged mainly along each side. The buds are tiny, almost hidden for most of the year. If a branch is lifted up, each of the needles will show off its bright white twin bands of stomata on the underside. Male flowers are carried densely on the side shoots all over the tree; they are small, globular, purplish-red at first, then yellow with shed pollen in April. Female flowers, produced on the same tree, are also numerous and terminate many short side shoots; eventually, these produce neat, little, nodding, pale brown, spruce-like cones about 2 centimetres ($\frac{3}{4}$ inch), long, each one having only about eight scales. It is unlikely that the tree has yet reached its ultimate size or age in Britain. It is severely affected on lime-rich soils by the root rotting fungus *Fomes annosus* and is not well suited to dry or chalky sites in eastern England.

Growth conditions

The western hemlock is hardy, but less so than the Sitka spruce or lodgepole pine. It will often survive on high moors and mountain sides, but is likely to be slow to establish itself and prone to producing multiple stems. It prefers areas of high precipitation and does best on well-drained sheltered sites that would otherwise grow Douglas fir and silver fir species. Frosty sites should be avoided because frosting of leading shoots in young trees will also cause multi-stem development. Hardly any tree is better adapted to shedding snow than the western hemlock; in addition, the flexible branches and narrow outline prevent windthrow unless rooting is very shallow or restricted. The western hemlock is a shade-tolerant tree, it retains its green needles for several years, and eventually shades out everything in the twilight beneath a continuous forest canopy. When plantations are eventually thinned out, allowing flecks of sunshine to penetrate to the forest floor, natural regeneration quickly occurs.

Propagation and management

From about the twentieth year of growth, good seed crops are produced. The cones and seeds are very small: a kilogram of clean seed may contain 200 000 individuals. Strong sunlight may damage nursery stock, and *Tsuga* was traditionally grown on a large scale in nurseries that were not in sunny areas, particularly in western Scotland. Provenance is important: those provenances that are prone to producing fluted stems or multiple leading shoots are now avoided. On suitable sites crop establishment presents few problems, although protection from deer and rabbits is advisable. Some side shading from ground vegetation is acceptable but swamping will cause foliage suppression. At thicket stage, it is worth spending a little time singling out (to one stem if necessary). The plantation will sustain itself at very close spacing with all the trees touching, and it is tempting not to thin them out. However, late thinning may cause the fragile, slender stems to snap, or the whole plantation to blow down, because the roots have been restricted.

Timber quality and uses

Western hemlock timber, widely used in America for building works, is a white wood which is not quite as pale or light in weight as spruce. It is creamy-brown with occasional brownish or grey streaks. It is non-resinous and straight-grained with mostly small knots and prominent annual rings. It dries rather slowly but is fairly stable in use, and takes a good finish. For

outdoor use it is weaker than Douglas fir and stronger than spruce timber, but is not durable and must be treated with preservative. A traditional use for small clean poles was ladder making. Western hemlock wood makes good pulp, but is not produced in large enough quantities to sustain a pulp industry in Britain. It can reach veneer quality for the manufacture of plywood. As a living tree its amenity value is low due to its sombre green look and dead twiggy appearance once the forest canopy has been broken. It is not suitable for urban areas and is intolerant of air pollution.

◀

The wood is white, non-resinous, and heavy when freshly cut.

17

Hornbeam

Carpinus betulus

L.

Hornbeam

Carpinus betulus L.

Carpinus comes from the Celtic *carr* meaning 'wood' and *pen* meaning 'head', recalling the old use of the wood to make yokes for teams of oxen. Hornbeam either means the same thing or, as Gerard suggested in 1633, a wood whose hardness may be compared with horn. *Betulus* refers to the superficial likeness to birch.

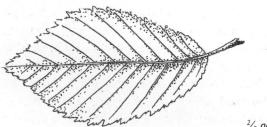

²/₃ *actual size*

Origin and distribution

In Britain, the hornbeam is only considered to be a native species in south-east England and in isolated areas in Somerset and Gwent. Its natural habitat extends throughout Europe, from the Pyrenees to southern Sweden and east to Asia Minor, and it has been introduced and naturalized over a very much wider area. In Britain there are records of introductions in the 15th century, at least as far north as Worcester and Norfolk. A few trees can now be seen in almost every part of lowland Britain, including the glens of Scotland. Hornbeam woods are rare, being largely confined to the areas of clay soil around London, but the tree frequently occurs as part of a mixed broadleaved woodland elsewhere. Isolated trees are frequently seen in the south-east as relics of wood pastures, an historically important source of the strong heavy timbers required by rural communities. These requirements are now either obsolete or can be met by using iron and steel. Hornbeam timber was often harvested from 2.5-metre (8-foot) pollards, leaving the stem, or bolling, to regenerate a new crop of branches, which allowed the land beneath to be used for grazing. Ancient pollard hornbeams still stand in parts of Essex and Hertfordshire, and in Hatfield Forest pollarding continues in the traditional way to this day, on trees up to 300 years old. Some of these trees have a stem diameter of 130 centimetres (13 feet 5 inches in girth). The tallest hornbeam on record in 1985 was in Cornwall at 30 metres (98 feet) tall. The largest diameter at breast height was measured in Hampshire and found to be 146 centimetres (15 feet in girth).

Male catkins, sometimes in great profusion, appear in early spring.
▼

Key features

The hornbeam is often confused with beech. Although the leaves are similar in size and colour, and have parallel veins, the margins, unlike beech, are sharply double-toothed, vaguely resembling those of birch. The bark is pale silvery-grey, in which pale brown fissures develop with age. Old trunks become deeply fluted and heavily buttressed, and invariably develop an oval cross-section. The heavy-branched crown is unevenly rounded and spreading, although freshly pollarded trees may develop a neat round head of branches for a few years before becoming irregular again. The shoots are dark brownish-grey with occasional silky hairs; they are thin and slightly angled between each bud. The alternate buds are sharply pointed and slender, like short beech buds, but tend to lie along the shoot rather than sticking out at an angle. The rough textured leaves are dark

▲
The green-winged seeds add density to the foliage in mid-summer.

▲
The seeds combine with the foliage to create a good display of autumn colour.

green with 9–13 pairs of pale, prominent, parallel veins. On the underside, each vein has a tuft of tiny white hairs at the point where it is joined to the midrib. The male catkins are very like those of birch except that they tend to be carried singly instead of in threes. The catkins appear as pendulous clusters of flowers, each consisting of a shared, light red outer bract and 12 yellow stamens. Although bunched tightly together, each of these in fact belong to three separate flowers. Female flowers are grouped towards the shoot tips where a cluster develops into a bunch of pale green, winged fruit in early summer. The seed wings have a unique and distinctive appearance: they consist of a 3–4-centimetre long tapered, papery bractiole with an enclosed flat seed at its base, flanked by two very small wings. The hornbeam is not known for rapid growth, but it does live for a very long time (over 300 years) and becomes an almost permanent landscape feature. In autumn, after a brief display of yellow and gold foliage colour, some trees will retain a few pale fawn-coloured dead leaves on their lower branches well into the winter.

Growth conditions

The hornbeam is essentially a lowland tree. It thrives on similar sites as beech, but is unlikely to progress as far up hillsides or grow as well as on thin, chalky soil. The best trees are found on rich loams and fertile clays. It is only hardy in southern England and sheltered microclimates further north, although established trees tolerate frost. The hornbeam grows well in partial shade and coexists with other broadleaved trees, to which it is frequently subordinate in mixed woodland. Well-rooted trees seldom blow over, but summer winds may dessicate the foliage on specimens. The seeds are a valuable source of autumn food for small birds, particularly tits and finches. Squirrels will take seeds but usually prefer larger, less labour-intensive fruits from other trees at this time of year. Woodmice and voles have plenty of time to clear up most of the seeds that fall to the ground, as seeds require 18 months to germinate.

▲
Ancient hornbeam pollards are still a common sight in south-east England. These are at Kew Gardens.

Propagation and management

The hornbeam is not widely managed at the present time, but there are several important features of management that are of particular interest. The nutlet is more difficult to clean than the seed of beech, chestnut, hazel, or oak. It is generally sown after only surface drying with the wings still attached as dewinging requires a special machine for large quantities. If the seed is collected early while it is still green, and sown immediately in the autumn before full dormancy has occurred, good germination should result the following spring. Fully ripe but dormant seed requires at least 18 months to germinate. There are about 24 000 clean seeds to the kilogram, which may be expected to yield 14 000 usable seedlings, but germination will probably be patchy and may be spread over three years. The traditional methods of managing hornbeam were usually pollarding or coppicing; the whole tree was seldom cut down and new trees were rarely planted. The hardness of the wood was notorious for taking the edge off of foresters' and carpenters' tools, so it is likely that the former avoided large fellings, and the latter had no need of huge timbers. The tree was a favourite species for 'wood pastures' rather than forests. In a wood pasture, individual trees were grown on open grazing land, usually as pollards, at very wide spacing. The trees had almost no harmful effect on grass quality and it appears that domestic animals and deer did not seriously damage the bark of well-established trees or roots.

Timber quality and uses

At one time, the hornbeam was the main source of very hard wood in Britain, usually in somewhat larger sizes than the equally prized boxwood. It was used to make mill cogs, piano parts, chopping blocks, and the like. These days it has only limited uses on a very small scale, mostly for high-quality craft items and first-rate firewood. Although the traditional uses for the wood have gone, the hornbeam is now highly valued as a living plant. It is an ideal hedge tree and, if trimmed in the summer, will hold a proportion of dead leaves through the winter and on to the following March. There are many cultivated forms of the hornbeam that are used as amenity or urban trees. The cultivar 'Fastigiata' is an ideal street tree because of its narrow crown spread, robust constitution, and fairly slow growth of 17 metres (56 feet) in 70 years.

▲
Hornbeam wood looks delightful but is impossibly hard to work.

18

European larch

Larix decidua
MILLER

European larch

Larix decidua MILLER

Larix is the Latin word for larch and *decidua* refers to the deciduous nature of the tree. The common name refers to the natural distribution of the tree, and recalls the German name, *Lärche*.

²/₃ actual size

◀
This 80-centimetre trunk was planted in 1919.

Origin and distribution

The European larch has a discontinuous natural range across Europe, but excludes Britain. There are four distinct areas of natural population although widespread planting has caused a degree of merging and confusion in mapping its range. Many seed lots from Alpine and Polish sources have produced very poor trees in Britain, much to the discredit of the whole species. Seeds of Carpathian origin are the best for use in Britain. The tallest specimen on record (1979) was 46 metres (151 feet). The largest diameter tree in 1991 was at Menzie Castle in Tayside, which had reached 192 centimetres (19 feet 9 inches in girth). The European larch was introduced to Britain before 1629, but extensive planting was not started until over 120 years later. The 3rd and 4th Dukes of Atholl found the tree to be well suited to their Perthshire estate, and two of the oldest specimens dating back to 1738 are found by Dunkeld Cathedral.

▲
European larch needles are soft and deciduous, turning gold before falling in late autumn.

◄
Spectacular pink, female flowers start to appear in late winter.

Key features

The European larch is an erect tree which has a tapering, usually straightish trunk with rugged, pink, brown, and grey fissured, scaling bark. The branches are short-lived and, in forest conditions, soon become exceedingly brittle and fall off, leaving a clear stem. Live twigs are pale yellowish-grey and glabrous. The deciduous needles are grass-green and linear, almost thread-like, arranged singularly and spirally on new growth, while old spur shoots have rosettes of up to 40 needles in a terminal cluster. In autumn the foliage turns from straw colour to old gold. The flowers are male or female, with both sexes on the same tree, even the same branch. Male flowers are globular and pale yellow, and females are upright and rosy-pink, like miniature cones with reflexed scales. Females mature in March some two weeks before the males on the same tree, thus avoiding detrimental self-pollination. The upward-pointing, leathery, oval cone is reddish in colour at first and then pale brown, with tough, thin, rounded scales. The small, winged seeds are carried two per scale. They ripen in one season and fall in the autumn. Old cones remain on the tree for many years, particularly on thin, pendulous twigs near to the ground.

◄
Numerous cones with soft, leathery scales occur on most mature trees.

Growth conditions

The European larch is not suitable for exposed positions, as trees there often develop a pronounced lean and one-sided heads of branches. Timber production in these circumstances is limited. The tree favours warm summers, a good water supply, and deep, moist soils which are slightly acid. In the lowlands, bracken is an indicator of a good larch site, whereas bramble is not. Late spring frosts determine where shapely trees are produced. The tree is deep-rooting on well-drained sites and is wind-firm if root spread has not been impeded. Full light is essential at all times, and larch forests benefit from heavy thinning from an early age. The ground under the larch is never quite bare of vegetation; below mature crops there is usually a continuous and healthy green grass sward. Being deciduous, the tree tolerates atmospheric pollution fairly well, but good specimens are seldom seen in the city.

Propagation and management

The seed conditions and nursery practice for the European larch are broadly similar to the Japanese larch. The European larch is fast-growing in early life, both in northern and southern Britain. Young plants often have semi-evergreen needles on their shoot tips. A good choice of provenance is important: trees from the Sudetenland mountains or further east in the Tatra region are favoured by foresters for their straight growth and resistance to larch canker, which seriously damages and deforms stems and branches of trees from many other European origins. The European larch is a valuable nurse tree on mineral soils, providing a continuous annual mulch of deciduous nutrient-rich needles for more valuable species planted alongside it. From this type of planting, or from closely spaced plantations, the small early thinnings provide valuable rustic poles. Thinnings should be marked in summer, when any trees which may be standing dead will not be overlooked. This light-demanding species must be spaced out so that branches in plantations hardly touch each other at any time after the first thinning.

Timber quality and uses

The resinous timber is a warm, reddish-brown or terracotta colour with orange and gold streaks, particularly towards the outside of the heartwood. The unevenly distributed sapwood is yellowish-brown. The heart of the tree is durable without artificial preservatives, and is very strong. This is an ideal tree for estate and agricultural purposes, particularly fence posts, gates, and rails. Traditionally, fishing boats were planked with European larch timber, and some of the wood still finds its way into quality boat building and repair work. It distorts unless dried very carefully, and the knots may fall out. Some individual trees have spiral grain, and may warp or spring when sawn. European larch timber has a tendency to split and thin boards are best drilled before nailing. Its durability makes it suitable for rustic garden screens and furniture, but it can be rough and quite crooked. The wood works and finishes well, but it requires skill to accommodate the knots, frequently changing grain, and resin pockets. If varnished the

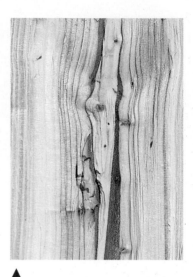

▲
European larch wood is valued for its strength, and its beauty is often overlooked.

effect is very pretty. As a living tree, the larch has some ornamental value, particularly in the autumn. Grown as part of a shelterbelt it has a stabilizing and wind-filtering effect, and will aid the growth of other species such as beech or pine. As a garden tree it is best grown as an isolated specimen. In full light, the branches can reach down to the ground, sometimes becoming massively horizontal with upwardly curving tips. In the spring and autumn it is outstanding for soft, subtle, yellow foliage colours.

19

Japanese larch

Larix kaempferi
(LAMB) CARR.

Japanese larch

Larix kaempferi (LAMB) CARR.

Larix is the Latin for larch tree and *kaempferi* commemorates the nineteenth-century traveller in Japan, Englebert Kaempfer.

½ actual size

◀ This tree is robust and hardy and will grow in harsh climatic conditions.

Origin and distribution

The larches are a small family of only ten species spread over the cool temperate areas of the northern hemisphere. The Japanese larch is a fast-growing, pioneer deciduous conifer species. It is native to the central Honshu area of Japan, from where it was introduced to Britain by John Gould Veitch in 1861. Its natural habitat is confined to Japan and includes Mount Fuji. All natural Japanese larch forests grow at altitudes above 1200 metres (4000 feet). In the British Isles this tree has been extensively planted in the west, particularly in Wales, but it is found in small patches almost everywhere. In Britain there are still some original and very early specimens of Japanese larch. After 1920, many conifer plantations were being established in Britain, and the Japanese larch was frequently used as a plantation edge species because of its ability to retard the progress of forest and heath fires. The tree itself does not burn easily, and its rapid early growth causes quick suppression of ground vegetation, which reduces available fuel in the path of a fire. By coincidence, using the larch in this way also disguises the sometimes dull green of evergreen conifer plantations. The limit of size here, of height at least, has just about been reached. The tallest recorded tree in 1989 was at Blair Castle, Tayside, standing at 40 metres (131 feet). The largest diameter to be found so far is at Dunkeld, where a tree in 1990 measured 105 centimetres (10 feet 10 inches in girth at breast height). Incidentally, these measurements are larger than any known in the original Japanese forests.

▲
The red shoots distinguish this tree from the European larch in winter.

The hybrid larch, *Larix × eurolepis*, is a cross between this species and the European larch. It was raised by the Duke of Atholl at Dunkeld House in Tayside. In 1885 he deliberately planted Japanese and European trees together, and after a few years collected seed from them. In 1904 the first obvious hybrids were identified by their intermediate shoot colour and enormous vigour. The hybrid larch has been the subject of much tree breeding since then, and experimental plots and small plantations of it can occasionally be found but they are difficult to recognize.

Key features

The Japanese larch is well known for its rapid early growth. Once established on a good site, young trees will grow to head height in about four years. They may then reach 20 metres (65 feet) in 30 years. Beyond that height, except on particularly favourable ground, the rate of growth will slow down. Tree tops may bend over or flatten out completely with age, and there is a corresponding reduction in annual ring width. One of the most striking visual features of Japanese larch plantations is that, in winter, the young twigs look rust-red in colour when seen from some distance away. Japanese larch woods stand out dramatically when lit by low winter sunshine in the morning or evening.

Close examination of a growing shoot reveals brownish hairs and a somewhat glaucous tinge. The soft deciduous foliage is of two kinds: on vigorous terminal shoots the 5-centimetre (2-inch) long needles are borne singly all around the shoot; on short spur shoots the needles are arranged in rosettes and are a little shorter. The first flush of pale blue-green foliage appears in mid-February. The needles mature to dark greyish-green until November, when

◄
The cones open out their reflexed scales.

they turn uniformly gold before falling to the ground. The flowers open with the leaves, males starting red-brown and becoming yellow, and females (which only appear on older shoots) being purplish-pink with reflexed scales. The flowers are numerous and may be found over the whole tree. The cones, which are ovoid, retain the female flower shape; the reflexed scales become papery and increase in size to about 3 centimetres (1 inch) and turn brown. Each cone has a short, curved, woody stalk which, regardless of branch angle, holds the cone in an upward direction. Japanese larch stems may be perfectly straight, or they may undulate as typically they snake skywards. Such stems are of limited value except for ornamental rustic work. The bark is always rough, but varies between individual trees, some being deeply fissured and others shedding numerous hard, curling flakes. The colour is uniformly reddish-brown, maturing to grey.

▲
Female flowers are spectacular in late winter and early spring.

Growth conditions

The Japanese larch is tolerant of most soils, but is more susceptible to drought than other larch species. It is hardy and wind-firm. As a pioneer species and colonizer of open ground, it will not tolerate shade of any kind. This may be the reason why, in crowded plantations, the rapid early growth period is very short. A huge quantity of dead needles fall each autumn, which smother and suppress most competing ground vegetation. This has been found to be enormously helpful to other, more tender tree species planted nearby, providing them with a regular supply of rich leaf litter as well as much needed shelter. Mature stands of larch, once thinned out, invariably encourage a ground flora dominated by a thick grass sward.

Propagation and management

Fertile seed can be obtained when the trees are between 15 and 20 years old. The seed is small and light, and one kilogram may contain 25 000 seeds. There may be 3–5 years between heavy cone crops, perhaps because the flowers are often damaged by spring frost. Cones must be gathered before they open, and somewhat earlier than those of the European larch. The gathering season extends from September in the south to October, or even November, in Scotland. The Japanese larch is easy to establish using standard forest planting techniques. Best results are achieved by using 2-year-old plants and planting them just before active growth begins in February. The trees soon become established and rapidly suppress any competing ground vegetation. Several defoliating insects attack the larch, one of the most serious being the larch sawfly, *Pristiphora erichsonii*. There is usually little point in maintaining larch plantations into old age, as productivity falls off dramatically after a very rapid start.

▲
The timber is not equal in quality to the European larch; small poles lack strength.

Timber quality and uses

Although the larch is generally known to be a valuable timber, the Japanese larch is the least durable of the whole genus. Tests have shown it to be less than half as strong as comparable samples of European larch timber. The heartwood is richly coloured red and brown, and the sapwood is paler, with conspicuous annual rings. There are numerous dark brown knots, which are often encased in bark, and tend to loosen in seasoning. Larch timber dries fairly rapidly but shrinks and warps badly. It is difficult to impregnate with preservatives, but takes paint and varnish very well. It is not particularly durable out of doors, although it is often used for rustic work, or sawn very thinly and treated with preservative for 'larch lap' garden fencing.

20

Small-leaved lime

Tilia cordata

MILL.

Small-leaved lime

Tilia cordata MILL.

Tilia is derived from the Greek *pteron* meaning 'feather' or 'wing', which refers to the distinctive bract at the back of the inflorescence. *Cordata* is from the Latin *cordatus* meaning 'heart-shaped', which describes the rounded basal lobes and pointed tip of the leaf. 'Lime' is probably a variant of the Old English name for the tree, *lind*, and the German *linde*.

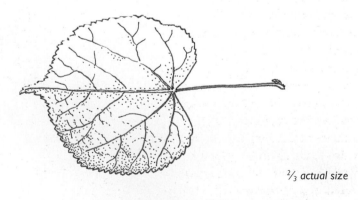

²/₃ *actual size*

Origin and distribution

In prehistory the small-leaved lime was the most common tree in lowland England, but climatic changes and grazing have taken their toll. By Anglo-Saxon times its prevalence was much reduced, and now it is confined to scattered, tiny strongholds, particularly in East Anglia, the south, and the Severn estuary. The small-leaved lime, or pry tree, should not be confused with the common lime (*Tilia × europaea*), which is confined to avenues and municipal parks, and is hardly ever seen as woodland. The common lime is a huge twiggy tree with bunches of whiskery shoots around its stem and often up the trunk. It is a hybrid between the large-leaved lime (*T. platyphyllos*) and the small-leaved lime, but is of uncertain origin. The more graceful small-leaved lime is native to Britain and most of Europe to the Caucasus, northwest Russia, and southern Sweden. In the British Isles, it is confined to England south of the Lake District and east of Devon, and to eastern Wales. Introductions have been made beyond this range, but seed is seldom produced and plants are always scarce. Most surviving lime woodland in England has been coppiced at some stage. Coppice stools are long-lived and some have been in existence for as long as 2000 years. The largest specimens known are in Wiltshire and Norfolk; the tallest was 40 metres (131 feet) in 1984 and the stoutest was 184 centimetres diameter (18 feet in girth) in 1991, at Haveningham Hall, Norfolk.

As coppice, small-leaved lime is almost immortal.
▼

Key features

The small-leaved lime grows to be a large tree of rounded outline with distinctive ascending main branches which are recognisable in winter from a distance. The outer shoots eventually turn downwards and those that reach the ground will layer. Frequent suckers are also produced from surface roots. Young trees, coppice shoots, and upper branches are coloured a dark, lustrous, olive-brown, and are sinuous. The shoot is uniformly brown all round with tiny, spaced-out, pale lenticels. The foliage is medium green and somewhat glabrous. The heart-shaped leaves are 4–7 centimetres (1½–2¾ inches) long with pale green backs. The petioles are slender and yellow, and as long as most leaves. In dry seasons the leaves tend to droop and curl, giving the whole tree, particularly when flowers or fruit are present, a pale yellowish, grey-green colour and dried-up look. The flowers appear in July. Each sweetly scented inflorescence has 5–10 yellow flowers on a slender stalk attached to a straw-coloured bract. The flowers on a tree are often numerous and very conspicuous. The thin-shelled spherical fruits, about the size of a lentil, are covered in yellowish-grey felt. Usually less than five fruits develop in each cluster, and most years they are infertile.

▲ The heart-shaped leaves vary in size from tree to tree.

▲ The seed ripens in hot summers, but is hardly ever produced in some areas.

Growth conditions

The natural range of this species, and its inability to set seed except in good summers, suggests that it is likely to be tender. In fact it has been found to be very difficult to kill, and stands up to East Anglian winter winds and frost, although it must be said it is nearly always in the mutual shelter of woodland. It is sensitive to grazing and can be shaded out by other tree species such as sycamore. The small-leaved lime appears to thrive best as a pure crop, as it will not compete with other to expand its territory. Lime, unlike hazel and (to some degree) chestnut, does not appear to suffer greatly when coppicing is abandoned. Very old stored coppice is often seen with few dead shoots, and little apparent root damage although stress cracks are common. Many good examples occur on rich calcareous sites.

Propagation and management

The small-leaved lime is not managed as a timber crop in Britain at present, although there is no reason why it should not be. It is recognized as an ancient woodland tree, and possibly one of the oldest living plants in Europe. Most important lime woods are in the hands of County Naturalists Trusts and conservation societies.

Lime coppices were formerly used for the production of underbark bast to manufacture coarse fibre and rope. The soft wood had numerous uses including charcoal. Management for production today has to be reconciled with the need for conservation. Propagation may be a problem. In addition to a scarcity of seed most years, the widespread planting of exotic limes and even *T. × europaea* can compromise the genetic purity of small-leaved lime seed. Modern vegetative propagation methods appear to be the best way forward. Planting should be done in the autumn, and weeds must not be allowed to compete at any stage of formative growth. Vermin, although not usually known to attack the lime, should be fenced out, as deer may graze new coppice growth. The use of treeshelters encourages rapid establishment of new plants.

▲
This ancient tree in Leigh Woods near Bristol could be 250 years old.

Timber quality and uses

In Roman times the small-leaved lime was known as 'the tree of a thousand uses', from household and agricultural implements to shields. It was said to be unlikely to be eaten by worms. The underbark or bast which, was particularly in demand, was soaked in water until it fermented a little, and then beaten to produce a coarse fibre to make strong rope, besom ties, fish nets, and rough clothing. On festive occasions head-dresses and ribands were made from it. Charcoal from lime wood was the best for artists, and was said to be excellent for gunpowder. The pale, straight-grained wood, which darkens on exposure to light, is ideal for shallow chip-carving—in the seventeenth century, Grinling Gibbons preferred lime for his work. More recent uses include toy making, pill boxes, hat blocks, bobbins, beehive frames, and some plywood. Lime wood does not taint and is suitable for dairy and domestic utensils.

The living tree has great potential as a woodland species and as an ornamental specimen. Avenues of small-leaved limes are more regular than seedling-grown common limes but, like the common lime, they can be attacked by aphids and shower down sticky honeydew in high summer. A splendid ornamental small-leaved lime is the cultivar 'Swedish Upright', a slender tree with level or drooping branches. Another narrow cultivar is 'Erecta', which has good yellow autumn colour and keeps its leaves until late in the year. Several very small-leaved forms are known in the wild, and the nursery trade are taking an interest in these. The small-leaved lime is a good bee tree, producing abundant nectar in July, which is not narcotic to the insects.

◀

The wood is soft and pale, and carves beautifully.

Field maple

Acer campestre L.

The common and scientific names have retained the form of the original Latin: *acer*, meaning 'sharp', referring to the use of wood from some maples for spears, and *campestre* meaning 'field'. A maple tree in Old English was *mapeltreow*.

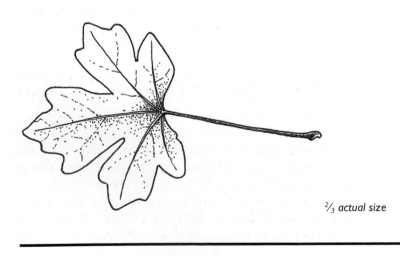

²⁄₃ actual size

◀

This is a tree of broadleaved woodlands where it is usually mixed with many other tree and shrub species in lowland areas.

Origin and distribution

The Mesolithic distribution of the field maple in Britain is uncertain because the tree sheds so little pollen that records are sparse. There can be no doubt, however, that field maple is a prehistoric tree in England and Wales. Medieval and Anglo-Saxon records indicate good numbers of field maple with a wide distribution. In this century, grubbing out of hedges and small woods has caused a dramatic decline in numbers of large specimens. However, the species has great powers of recovery, will coppice freely, has been widely planted and has become naturalized. It is not threatened, and numbers have been increasing since about 1980. Beyond the British Isles, its natural range extends across Europe to Russia, Asia Minor, and southwards to North Africa. Although usually a component of mixed woods, notably with ash, hazel, and oak, remnants of ancient, pure maple woods can be found, on soil over limestone or chalk. The field maple is an inherent part of Wychwood, Hatfield, and Rockingham Forests, and can still be found on the North Downs near Box Hill, where maple woods were once known to be extensive. The tallest British tree on record is 27 metres (88 feet). The largest stem diameter in 1992 in Essex was a pollarded tree of 141 centimetres (14 feet 6 inches in girth).

Key features

The field maple is not naturally a large tree and measurements half those of the existing champions are considered to be good. The tree has a domed crown when growing in full light in the open, which is carried on a trunk which is sometimes whiskery and burred, with short side shoots all the way down. Young bark is soft and corky, pale brown, uneven and sometimes having extended vertical ribs of cork on 4–6-year-old wood. The bark becomes rough and harder with age, and fades and weathers to pale grey. Young shoots are deep red-brown, usually with a lighter shaded side. They are finely pubescent at first. Trimmed field maple as part of a hedgerow will produce distinctive red young foliage, and then rich, red winter twigs.

▲
Rich, green field maple foliage is both shelter and food for numerous insects and other wildlife.

▲
The blushed seeds are especially beautiful in August.

The buds are also reddish with a grey inward side. The leaves are small and palmate, with five wedge-shaped lobes; they are downy at first, deep green above and slightly paler below. Late in the summer, the foliage is often dusted with grey mildew. The petioles are slender, sometimes quite long, green, sometimes with bright pink on one side. Prominent yellow and pink autumn colour can be relied upon, and individual trees may have a tinge of orange. The flowers are greenish-yellow, about 7 millimetres (¼ inch) across, with maybe ten in each erect inflorescence. The flowers are usually bisexual, but sometimes a tree will produce only males. Perfect flowers have narrow petals and sepals dominated by eight yellow stamens. Fruits develop from about three of the flowers in each inflorescence, each one consisting of a joined pair of seeds with asymmetrical, stiff, membranous wings held almost horizontally opposite each other. By late summer these pale green wings may be flushed with crimson before ripening to brown. Growth can be rapid at first on good, damp sites but it usually settles down to a steady slow rate. These trees can be long lived (300–500 years), coppice and layered hedges perhaps even longer, but severely damaged trunks are not durable and rot quickly sets in. After an initial year of straightish growth, coppice soon becomes a tangle of twisted, wiry shoots which have little use.

Growth conditions

Fertile, heavy, moist, calcareous soils are required for successful establishment and good growth of field maples. On chalk, the roots appear to be able to draw enough water from the rock to sustain the tree on what may seem to be a very dry site. The British climate does not seem limiting in terms of survival, but tree quality is likely to be inferior on frosty, windy, and high elevation sites (in any case, suitable soils are not common in such places). Being small and rounded, the field maple seldom blows down, although it is often smashed by neighbour-

ing larger trees of other species as they fall. The field maple is a good bee tree and provides shelter and food for a large number of insects and some small mammals. Birds make good use of its tangled branches and gnarled stem for nesting.

▲
Field maple is very much at home as an isolated hedgerow specimen.

Propagation and management

As a general rule, seed should be sown as soon as it is ripe in the autumn. If practicable, the wings should be removed. Special precautions must be taken against damage by mice. Dry seed should be stratified for one year before sowing, but this chore can be avoided by collecting seed in September before it dries out. Plants will require at least two years in the nursery bed or container to reach plantable size. Planting on sheltered sites is best done in the autumn, and protection from rabbits in particular is vital. An area round the plant must remain weed-free, as lush grass and weeds will outgrow newly planted field maples and swamp them if not restrained. Although many insects eat the leaves, there are few entomological problems or diseases specific to this tree.

Timber quality and uses

The timber is fine-grained and pale. When cut from the tree roots or from a burr, it is often beautifully veined. From behind a cluster of epicormic shoots it will have the highly valued 'bird's eye maple' figure. Maple wood has been used for cabinet making, musical instruments, and all kinds of high-quality woodwork in the past, but tends not to be in demand today simply because trees are not available to be cut down. As a living tree the field maple has much to commend it. It is a native species suitable for planting in association with several other British native species. It has a part to play in the food chain of many organisms, so aiding their survival and conservation. It is a small tree that is good for confined spaces, it casts only light shade, and has glorious autumn colour.

▲
British-grown native maple timber once had numerous uses, especially specimens with wavy grain and burrs with a 'bird's eye' figure, in fine furniture and musical instruments.

22

Pedunculate (English) oak

Quercus robur

L.

Pedunculate (English) oak

Quercus robur L.

Quercus is the classical name for oaks, said to be from the Celtic words for 'fine' and 'tree'. *Robur* describes in Latin all hardwood, especially that of oak. The name 'oak' is from Anglo-Saxon *ak* or *aik*, the grain or seed being called '*aik-corn*'. 'Pedunculate' refers to the fact that, in this species, the acorn cup is attached by a stalk or peduncule.

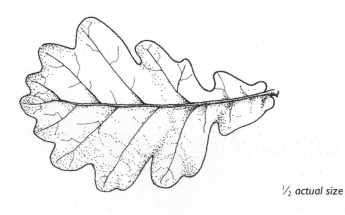

½ actual size

▲
Acorn cups of English oak have slender stalks.

Origin and distribution

The pedunculate, English, or common oak is one of the most familiar broadleaved trees in Britain. Only willow, birch, and alder are more widely distributed throughout the British Isles. The natural range of the species includes the whole of Europe to the Caucasus, northwards to Scandinavia, and south to Asia Minor and North Africa. On fertile, lowland sites it either forms massive isolated trees or tall dense forests. On high ground it is more often seen as low twisted and stunted woodland. Wistman's Wood, high on Dartmoor, is a good example; the trees there were measured in 1621, and their height is exactly the same today. The oak has featured strongly in the history of Britain, and consequently numerous notable old trees have been measured and recorded. One of the oldest specimens was a 'bog oak' unearthed from the peat on Stilton Fen in 1948 after being buried for thousands of years. Its height was 46 metres (151 feet) and its stem diameter was 136 centimetres (14 feet in girth). In 1796, a huge living tree in Wales was measured at 21 metres (70 feet) to the first branch and 204 centimetres across the base (21 feet in girth). The tallest living trees known today are 42 metres (138 feet). The diameter of the largest surviving hollow and stunted oak in Britain is 384 centimetres (39 feet 6 inches in girth). It could be 1400 years old, and is still alive.

Native oaks in Britain belong to two different species. The 'English oak' has stalked acorn cups. It is a very common tree which is distributed throughout the whole country, although it is confined to low ground in mountainous regions. Sessile or durmast oak, on the other hand, has stalkless acorn cups and prefers damp rocky hillsides and siliceous gravelly sites. Mixed woods of the two species do occur, and there may be many confusing intermediate forms, but true hybrids in the botanical sense are infrequent. Another oak frequently found in British woods is the naturalized turkey oak, *Quercus cerris*. This vigorous tree has a typical oak form but the dark leaves are longer, narrow, and have more lobes. Their best identification feature is the coarse curly bristles on the outside of the acorn cup.

Key features

The pedunculate oak is a deciduous tree that may live for a very long time. It progresses from a young, smooth, silvery-brown barked sapling to a huge, rugged, hollow hulk, with rough, hard, deeply fissured bark. The typical development of the tree includes a period of quite rapid early growth for around 80–120 years, followed by a gradual slowing down. After about

250–350 years decline sets in, branches die back, and diameter growth slows right down. Trees surviving beyond 400 years will have gradually replaced most of their heavy branches with a shorter head of branches in keeping with their more modest requirements, a kind of natural pollarding. Oak leaves are quite variable in outline, normally up to 10 centimetres (4 inches) long with about six pairs of rounded irregular lobes cut into their margins. The leaf base tends to taper slightly, ending in a pair of uneven glandular lobes and the stalks are stout, curved, and short. The upper surface of the leaves at first appear tinged with brownish-red, then becoming dark green, and the leaf back is grey-green and almost glabrous. In late summer, new red 'lammas' foliage appears, which compensates for earlier losses to insect predators—Lammas Day itself is the 1st of August. In late autumn oak leaves turn to old gold and rich brown before falling in November. Young trees in particular sometimes hold a few dead leaves through the winter. The clustered buds are pale brown with silvery fringed scales. The largest terminal buds are somewhat pentagonal in cross-section. Catkins open in May along with the new leaves. Male flowers are on pendulous tassels, and the females appear as tiny, pink, bud-like structures on the shoot tips. One to three acorns are formed on a slender stalk, each being contained in a cup about one-quarter of its own length. Good seed, or mast, years occur every 4–6 years.

▲
Mature bark is rugged and strong.

▲
Oak flowers appear in spring with the unfolding leaves.

Growth conditions

The pedunculate oak is very hardy, and only very young plants are affected by late spring frosts. It grows on most fertile soils, but in Britain the timber quality varies from one locality to another, particularly in areas where free-draining land and clays meet. Freely draining, stony, and gravelly soils are associated with poor oak affected by timber 'shake' and longitudinal cracks which reduce the value of the logs. The best oak is often grown on heavy, moist, clay soils. Some salinity is tolerated, so occasional flooding by brackish water is endured without much damage. Rooting is deep and wide-spreading, so windthrow is not usual. Young plants can tolerate shade, but the trees strive to reach upward and out to full light. Healthy ground vegetation, including brambles, bracken, and bluebells can survive under oakwood. Woodlands that have had a sustained phase of management such as broadleaved mixed coppice may be havens for wildwood flowers such as oxslip, wood anemone, herb Paris, and lily-of-the-valley. Many such coppices may have an element, or a majority, of oak in them, especially as 'standards'. New oak woodland that is not artificially planted usually develops from neglected grassland or from broadleaved scrub, provided that there are seed trees nearby. Dominant specimens assert themselves quickly, although uncompetitive subordinates grow tenaciously alongside. Trees with frequent burrs and ridges on their trunks are likely to live the longest because they have the ability to produce new boughs late in life, but they are less valued by the forester for timber. The survival of British oak woods, prior to widespread planting, has depended almost entirely upon the jay, *Garrulus glandarius* because acorns taken away from the shade of the parent tree and buried by jays are the only ones with much chance of survival. Squirrels and other acorn-hoarding animals invariably damage their prize in the process of storing it. In recent years most acorn crops have been severely reduced by the spectacular knopper gall wasp, *Andricus quercuscalicis*; fortunately the insect does not affect the health of established trees.

Propagation and management

A mature oak in a good seed year will produce 50 000 acorns. Although they vary in size and weight, about 300 acorns may be expected per kilogram. Seedlings grow rapidly and produce long taproots, which in the nursery are usually pruned back or 'under-cut'. Planting oak trees presents few problems but, because fertile sites are usually chosen, weed competition may be severe. Plastic treeshelters can safely be used to speed up early establishment, they also provide protection from rabbits, voles, hares and, to some extent, deer. Numerous defoliators live on the oak, but they are only a problem for the forester or for the tree if they reach epidemic proportions. Some incidence of epicormic branching (tufts of shoots growing out of a mature trunk) will be found in most oak woods. The most affected trees are usually thinned out of commercial plantations. In the past, small-diameter oak wood was taken from coppices. The woodland was cut over every 20 years or so and the stumps allowed to regrow. Traditional management included 'singling' to about three shoots per stool at four years (waste-weeding), and thinning again in mid-rotation for charcoal and firewood (cordwood-weeding). Stools or stumps of managed oak coppice have an almost indefinite life.

Timber quality and uses

Historically, the oak produced four essential products: tanbark, timber, mast for animal feed, and small diameter wood. Tanbark would be stripped from trees, felled or standing, in April, May, and June; this practice virtually ended in about 1880 when vegetable tanstuffs were imported from overseas. The timber industry continues, but with modifications. In 1860 the

◀
Knees and ribs in old wooden shipbuilding.

▲
Near quarter-sawn (through the centre of the tree) oak wood.

Battle of Hampton Roads proved decisively that iron ships were superior to those of oak, ending the construction of large wooden ships a tradition many hundreds of years old. Oak for structural work, particularly restoration, is still very much in demand. The 'small oak wood' industry provided charcoal, firewood, and pitwood; in addition, small trees were squared up with an adze for structural work before saws were perfected. None of these demands on oak remain today. There are, however, still numerous uses to which oak is put: the durable heartwood is good for estate work, gates, and fence posts; well-grown oak is eagerly used by craftsmen and quality furniture manufacturers; the best oak is used to make veneers and barrel staves which impart tannin flavours to spirits and fine wines; there is still a market for heavy, sawn-oak timber, wooden sea defences, and firewood; and hardwood pulp can still be made from oak. The living oak tree has high amenity value, old trees being particularly cherished. The oak is frequently chosen for ceremonial planting, giving a degree of permanence that few other species can match.

23

Red oak

Quercus rubra

L.

Red oak

Quercus rubra L.

Quercus is the classical name for oaks, said to come from the Celtic word meaning 'fine' and 'tree'. *Rubra* means 'red', from the Latin *ruber*, and describes the autumn foliage colour.

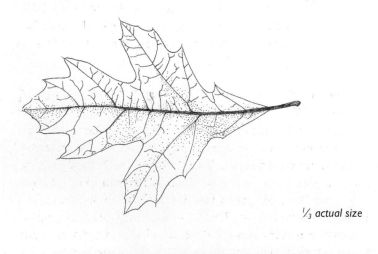

⅓ actual size

▲

In lowland Britain, the red oak usually grows faster than the native oaks.

Origin and distribution

This large tree is one of a number of fairly similar-looking, closely related American oak species. It is probably the largest of the group and perhaps the stoutest, and has a natural range extending through Quebec, eastern Canada, and the eastern United States. In the south it can be found as far west as Texas. Its date of introduction to Britain is somewhat uncertain; 1724 has been suggested, and it is known that a batch of seed from North Carolina was brought in in 1730. Until the 1940s it was hardly thought of as a forest tree at all, but it was widely used to enhance ornamental collections and amenity belts in parks and gardens. The red oak can be seen now in this situation on many estates, either as a stately individual or as a clump or roundel mixed with the pin oak *Quercus palustris* or scarlet oak *Quercus coccinea*. Many small trial forest plots have been established, and the species has often been used as an ornamental edge to plantations of other less attractive trees. The red oak has been widely planted throughout the whole of Britain, particularly as an urban park tree. The best examples known are at Westonbirt, which had the tallest (32 metres (105 feet) in 1988), and Cobham Park, which boasted a tree 202 centimetres in diameter (20 feet 10 inches in girth) in 1990.

Key features

The red oak has thin bark which resembles beech bark for most of its life. It is usually shining silver grey and fairly smooth. Young stems and branches are slightly browner than beech but become silvery later in life. Some trunks fissure vertically with age, others turn brownish-grey. The young shoot is dark brown, ribbed, lustrous, and hairless. The buds are similar or darker in colour to the shoots, ovoid, and pointed. The matt green leaves are variable in outline but always quite large, up to 20 centimetres (8 inches) long. There are 4–5 large sharply angled lobes on each side, which are to a greater or lesser extent further divided and pointed, each point being terminated by a distinctive whisker. The depth to which the gaps between the lobes cut into the leaf is variable; leaves with many intermediate lobes appear to be the most deeply cut. The male catkins are slender, pendulous, and up to 8 centimetres (3 inches) long. Female catkins, which appear in the leaf axils of new shoots, are small, globular, and dark red. The acorns take two seasons to reach maturity, with shallow cups and scales of pinkish-brown with purple margins. The growth rate is usually somewhat similar or a little faster than the pedunculate oak, but growth rates of up to 2 metres (6 feet 6 inches) a year have been claimed for a short period on young trees. The life span is not particularly long in Britain and trees over 150 years old are usually found to be in an advanced state of decay. In its native habitat, the red oak lives much longer than this.

▲
The catkins appear in spring with the leaves.

Forests are unusual, and at this stage in winter they resemble beech.
▼

Growth conditions

The red oak has been tried as experimental plots on most kinds of soil. It prefers good lowland loams for unhindered development to great size, but has shown a remarkable ability to survive on more severe semi-upland sites. On dry, sandy soils it will usually survive to average tree size where the English oak might not be more than stunted shrub. On acid, heather-covered uplands, young plants will hardly grow at all, but given shelter on such sites, particularly when nursed by larch trees, they soon show some promise. In addition to producing tree cover, the trees gradually improve and enrich impoverished soils with their substantial roots and broad deciduous fallen leaves. Although the red oak is damaged by severe cold or drying winds, it is only usually completely killed by exposure when it is planted in absolute isolation. Older trees are seldom blown down, but in old age serious crown damage is often caused when heavy weakened branches snap off and crash through the canopy. Full light is required for best growth but, provided there is enough top light, trees grow well in association with each other.

Propagation and management

Very little management of the red oak has been done in Britain. Good seed is only occasionally produced here, so most seed has to be imported from America. Nursery practice is then much the same as for other oaks. The number of acorns to a kilogram is likely to be in the region of 280, but degrees of drying during importation will cause this figure, and possibly the viability, to fluctuate. Larger acorns tend to be produced from more southern areas. The seedlings grow quickly but are less likely to produce such dominant taproots in their first year as the pedunculate oak. Hot sun on the young foliage may cause scorching in the seed bed, and shading should be used in the same way as for beech. When planting out, plastic treeshelters can be used to aid establishment; half-height shelters (about 60 centimetres tall) have proved very suitable. Although the red oak is attacked occasionally by some of the defoliators of British oaks, it usually manages to keep most of its leaves intact until they fall in the autumn. There are no traditional methods of managing this species in Britain. In America, red oak timber comes increasingly from regrowth following the extensive clearances of the nineteenth century.

Timber quality and uses

Almost nothing is known about the quality of British grown red oak timber. Only occasional parcels of thinnings or the old big estate tree come on to the market. These generally find mundane uses such as box making, pallets or firewood, and are never included with quality European oak, which is at least 10 per cent lighter in weight. Although the pinkish-brown wood of the red oak is heavy and hard, it is not particularly decorative. If supplies in Britain were more plentiful, it is possible that an industry would build up in response. There is increasing interest in planting the red oak in France, Holland, Belgium, and Germany. So far it has no particular uses in Europe, but it was at one time widely used in America for structural work.

► The wood is at present mostly used for low-grade work such as pallets.

▲ British-grown timber is hard and strong.

In areas where oak woodland was dominant, it must have provided most of the material to build and run whole communities. The living tree has great amenity value, is deciduous, and will tolerate forest sites where it is usually used to edge conifer plantations. It is wind-firm and helps to improve the stability of such crops, and will tolerate industrial sites. In autumn, its foliage can be a good, uniform, ruby-red colour, although it cannot always be relied upon to be so, and may just turn dull brown. In spring, the newly flushed foliage is always pale yellowish-green for about three weeks, and in summer the large leaves provide good shade. It makes a fine, but somewhat large, town tree. In isolation it often forks low down and spreads widely, but in a group it will produce more stately longer boles.

24

Sessile oak

Quercus petraea

(MATTUSCHKA) LIEBLEIM

Sessile oak

Quercus petraea (MATTUSCHKA) LIEBLEIM

Quercus is the classical name for oak. *Petraea* is from the Greek *petra* meaning 'rock', a reference to this tree's preference for stony places. Sessile means 'stalkless' from the Latin *sessilis* and refers to the stalkless acorns. Oak is from the Anglo-Saxon *aik*.

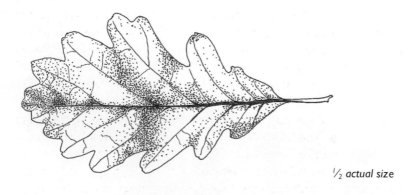

½ actual size

◄

Sessile oaks reach into the mountains of western Scotland.

Origin and distribution

The sessile oak is native to central and western Europe, including Britain, and its range extends as far east as Asia Minor. In Britain it is the dominant species of oakwoods in the north and west and may occasionally be found intermixed with pedunculate oak in the south. A sessile oak was the tallest British oak on record, having reached 43 metres (141 feet) in 1984. The largest stem diameter on record is 357 centimetres (36 feet 10 inches in girth at breast height) at Croft Castle in Herefordshire. This tree is believed to be over 1200 years old, and has been pollarded many times.

Key features

The sessile oak is distinguished from pedunculate oak by its rather larger leaves, which are usually cuneate towards the base and not auricled. The leaf stalks are longer and are usually of a conspicuous yellow colour. Leaves of all sizes and shapes can, however, be found on both species and some individuals conform to the descriptions rather better than others. The acorns, when present, are stalkless. Upswept branches are said to be a feature of the sessile oak, but this too is unreliable; some pedunculate oaks also have them. The foliage is dark green through the summer, becoming paler and yellowing in the autumn. The rather thick leathery leaves are ovate with about six pairs of rounded lobes; the margins are entire but the outline is variable. The underside of the leaf is pale green, and there is usually a certain amount of pubescence on the mid-rib and main veins. The young shoot is dark purplish-grey, usually with some pale bloom on it. The prominent winter buds, with many pubescent pale orange-brown scales, are clustered on the shoot tips. The branches are light grey and smooth at first, but become finely fissured with age. The trunk is often straight and persists through the centre of the crown at least until decline sets in, usually at an age of about 300 years. The bark is similar to that of English oak: rough and hard, vertically ridged, and light grey in colour. Numerous drooping male catkins appear with the new leaves in May after about the age of 20 years. Female catkins are borne near the tips of the sideshoots, each with a tiny reddish-purple

▲
Probably the safest identification feature is the virtually stalkless acorn cup.

▲
An old tree on a rocky site in Wales.

stigma held close to the stem. The acorns, which mature in one season, are short and bluntly conical in outline and are held in stalkless cups. The sessile oak lives for a very long time, possibly over 1000 years, but there appear to be fewer surviving old hulks around the countryside than of the pedunculate oak. Many of the oldest trees appear to have been pollarded at some time in their history.

Growth conditions

The sessile oak prefers damp rocky hillsides, and is most at home in western Britain. It is totally hardy throughout the whole of the country and prefers soils just on the acid side of neutral. If it is transplanted to a lime-rich, fertile, lowland site, however, it appears to adapt very well. Small plants cannot withstand competition from vigorous ground vegetation and may simply refuse to grow for many years, but plastic treeshelters speed up early growth considerably. This tree is one of the most wind-firm in Britain, especially on rocky sites. It is tolerant of cold and, should a spring frost damage the foliage, a new flush of growth ensures a quick recovery. Maturing trees must have full light at least near the crown. There seems to be fewer foliage predators than on the pedunculate oak. Even the common defoliating oak roller moth, *Tortrix viridana* usually prefers the pedunculate oak.

Propagation and management

There are some 300 acorns in a kilogram of seed, giving up to 250 usable plants after one year. In the nursery, if plants are not lifted or undercut each year, they form a strong deep root system which makes subsequent plant handling difficult. Planting sites should remain free of weeds next to the tree, and vermin must be controlled for as long as possible. On poor sites the species responds well to the nursing effect of larch trees in early life. Small thinnings are of limited value, and there has been a tendency recently to plant at wide spacings, or to grow the oak in combination with short-lived species. However, the variability of oak and its frequent deformities exposes such practices to risk, leaving too few trees from which to select a good final crop. Timber can be produced in as little as 80 years, but forest rotations of 150–200 years are traditional.

▲
Young plantations are traditionally planted close together to draw up clean stems.

Timber quality and uses

Sessile oak timber is similar in most respects to English oak timber. It is hard and decorative with a particularly attractive figure in radial cleft or quarter-sawn material. The heartwood is durable and on the whole straighter than that of the pedunculate oak. It has many uses on estates, particularly for gates, gateposts, split rails, fence posts, and strainers. Evenly grown material is suitable for high-quality furniture, panelling, and veneers. Barrel staves from the sessile oak forests of Limousin and Troncais impart a particular flavour to brandy, whisky, port, and sherry. In the past, most sessile oak woods were coppiced or pollarded for tan bark, charcoal, small building timbers, and fencing. The waste material provided good-quality firewood. Sessile oak timber is rather difficult to nail and reacts chemically with iron, so in damp conditions a dark purplish-black stain spreads through the wood where it is in contact with any ferrous metal. The traditional use for shipbuilding continues to this day, albeit now on a rather modest scale.

▲
The quality and strength of sessile oak timber can be outstanding.

25

Corsican Pine

Pinus nigra var. *maritima*
(AITON) MELVILLE

Corsican Pine

Pinus nigra var. *maritima* (AITON) MELVILLE

(*Pinus nigra* subsp. *laricio* (Poir.) Maire, is the accepted scientific name for this tree but var. *maritima* is still widely used by foresters in Britain). *Pinus* comes from the Latin for the pine tree. *Nigra* means 'black' and refers to the stems. *Maritima* refers to the coastal nature of this race of the black pine, *laricio* is the Italian name for several pines, possibly from *larici* meaning 'larch-like'. Corsican refers to an area where this tree grows.

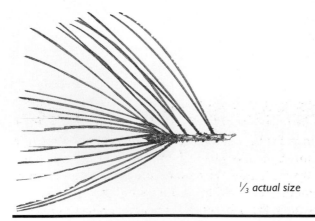

¹⁄₃ actual size

Origin and distribution

This is the most frequently planted introduced forest tree in southern England. It is a
variety of black pine, which occurs naturally
only in Corsica, southern Italy, and Sicily.
Other varieties of *Pinus nigra* are found
throughout southern Europe from Spain to
Turkey, and there are also isolated populations in northern Morocco, Austria, and the
Crimea. These geographical races tend to
integrate with each other, and precise
identification of each one out of its native
habitat is difficult. Although planted widely
throughout the British Isles, particularly from
1950–70, this tree is only at its best in the
south. Hot, dry, sandy soils are preferred, particularly at Wareham, the New Forest, and
Thetford. Plantations north of Leeds, except
where there is a mild microclimate, tend to be
sluggish in growth and prone to disease, particularly *Brunchorstia pinea*. The introduction
date to Britain for the species and the variety
is unclear, but in 1759 an account of a tree
thought to be Corsican pine appeared in the
seventh edition of *Miller's Dictionary*. The
Corsican pine was not widely available,
however, until the 1820s. A tree at Kew
known to have been planted in 1814 is said to
be the oldest surviving specimen in Britain.
The tallest British tree recorded is 46 metres
(151 feet) and the largest stem diameter in
1991 was 179 centimetres (18 feet 6 inches in
girth). In France trees of such size are usually
thought to be in excess of 360 years old.

▲
This 100-year-old tree shows typical form and colours
of bark.

Key features

Of all the varieties of black pine the Corsican is the best and most convenient timber tree; the
stems are straight, usually monopodial, and mostly free of heavy branches. The branches grow
in distinct whorls, often very evenly spaced up the trunk. The bark becomes dark grey, rough,
and shaggy after a few brief years of being smooth. Very old trees show smoother flat scales of
pale-grey and pinkish-brown. The foliage is dark greyish-green. The current shoots are pale

▶
Clusters of male flowers shed huge amounts of pollen in the spring.

▶
The minute female flowers are seldom seen.

yellow-brown for one year, older shoots being dark brown and rough. Needles are carried in pairs, and are spreading and sometimes twisted. They are longer than other common pines in Britain, up to 18 centimetres (7 inches), with a persistent grey basal sheath. Green foliage and lower branches are quickly shaded out in forest conditions, and good, clean, knot-free timbers are produced as a result. Winter buds are narrowly conic with a sharp pointed tip. Basal scales are usually distinct, but most of the dark brown bud is usually encrusted with white or translucent resin. Male flowers are clustered towards the base of current side shoots; they are purplish to begin with, then dusted with abundant yellow pollen grains in late May. Female flowers are pinkish, and occur on the tips of expanding shoots, developing in one season into conelets, and then in the second year to full-sized, shining, pale brown, woody cones 6–7 centimetres (2.5 inches) long. In its warm natural habitat this is a long lived tree, but it does not appear to live much more than 150 years in Britain. Nevertheless ultimate sizes are about equal in Britain and Corsica.

► The woody cones take two years to ripen and are freely produced.

Growth conditions

Of the pines described in this book the Corsican pine is the one likely to survive best in alkaline soil conditions. Various black pine varieties are used for garden, park, and sea-side windbreaks; they are resistant to wind and salt spray, and provide good shelter. Although the tree will grow in favoured areas throughout the British Isles, this country is a long way north of its native habitat and it will not flourish where summers are cool. The Corsican pine is a resinous tree, and in hot dry periods it may be prone to destruction by forest fires.

Propagation and management

One difficulty in establishing this tree lies in its root system: fibrous roots are not freely produced on seedlings, and transplanting bare-rooted stock can involve heavy losses. To overcome this, containerized plants can be used. These are best planted out in February when transpiration is low and the ground is usually damp. Once established, the tree's long probing roots are put to good use, enabling it to withstand spring drought and low summer rainfall, particularly on dry sandy soils. When fertile lowland sites are planted with Corsican pine, competition from weeds is not tolerated until thicket stage. Broadleaved woody weeds such as bramble, birch, sallow willow, or chestnut sometimes become serious competitors and threaten the survival, and form of the pine. At pole stage this light-demanding species must still be given space. Failure to thin out plantations early enough can cause suppressed trees to die and others to become very slender and unable to support themselves later on. A watchful eye must be kept for signs of disease at all times, and cut stumps should be treated immediately after felling to prevent infection by fungal disease such as *Fomes annosus* (butt rot).

Timber quality and uses

The timber quality and straightness of Corsican pine is superior to all the other black pine races. The wood is hard, strong and resinous, and there are prominent resin ducts and distinct annual rings. Old trees can produce long lengths of knot-free timber but the inferior sapwood tends to be wide or variable in extent. The heartwood is reddish and the sapwood is pale yellow. The Corsican pine is suitable for most of the purposes to which the Scots pine is normally put and it is a good structural and joinery timber. At one time, resin was obtained from standing trees by tapping, but the quantity and quality of the resin was always less than that of maritime pine (*Pinus pinaster*).

◄

The timber is sometimes knotty and resinous, but relatively strong.

26

Lodgepole pine

Pinus contorta

LOUD.

Lodgepole pine

Pinus contorta LOUD.

Pinus is the Latin word for all pine trees, and *contorta* refers to the contorted nature of the stem, branches, and needles, particularly of the subspecies *contorta*, the coastal form. Lodgepole refers to the use by American Indians of the slender stems of subspecies *latifolia* (Engel.) Critchfield, to support their lodges. Pine comes from the old English *pin*, from the Latin *pinus*.

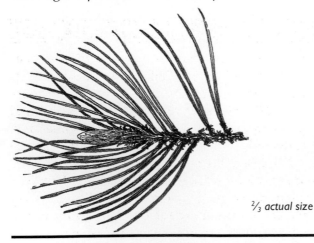

²/₃ *actual size*

◄

Some inland provenances make straight stems.

Origin and distribution

The lodgepole pine has a wide natural range in north-west America. In the north it extends from the Yukon to the southern Alaskan coast, through British Columbia and Alberta. Its eastern limits are the ramparts of the Rocky Mountains as far south as The Cascades and Sierra Nevada, where it can be found at an altitude of 3350 metres (11 000 feet). It was discovered twice by Europeans, and its introduction to Britain was not straightforward. The inland form, *Pinus contorta* subspecies *latifolia*, was discovered by John Jeffrey in 1805, but it was the 1825 discovery of the coastal form (beach pine) by David Douglas, near the mouth of the Columbia River in Washington State, that appears to have been introduced in 1831 and later lost. Seed from Jeffrey's original discovery was finally introduced in 1854, and the Douglas discovery was then reintroduced in 1855. The inland and coastal varieties have been widely planted in British upland forests. The species in its natural state is so varied that care is needed to select a suitable provenance or seed origin for a particular site and purpose, guidelines for which have been published. Vigorous coastal forms planted on soft wet ground are sometimes unstable at first and will bend; inland forms are more wind-firm and straight, but some may be particularly slow-growing and prone to disease. Good forms of both inland and coastal trees have now been selected for use in the British Isles. The tallest tree in Britain in 1989 was of inland provenance planted in North Wales, recorded at 34 metres (111 feet). The largest stem diameter known in 1986 was 114 centimetres (11 feet 9 inches in girth).

Male flowers are often produced in profusion after six years.

Key features

It is almost impossible to identify this tree from its outline alone as it varies from a low contorted shrub to a slender columnar tree. It commonly tends to have quite short branches. The bark is drab grey-brown and breaks into small, rectangular, solid plates with vertical fissures on stems over about 15 centimetres (6 inches) in diameter. Young twigs are green at first, then turning brown. Winter buds are strongly resinous and blunt-ended. The needles occur in pairs, with a short-lived, dark grey, papery basal sheath, and are shining mid-green, often slightly twisted and tending to be yellowish-green at the base. Needle length varies according to

Female flowers are small but brightly coloured.

type and may be 3–8 centimetres (around 1–3 inches). Male flowers appear in numerous clusters of 20 or more, midway along expanding current sideshoots. Female flowers occur on the upper expanding shoot tips as spiky conelets about the size of a pea in the first year. They develop into full-sized 5-centimetre (2-inch) woody cones. They always have short stalks and tend to point backwards along the branch. Each cone scale terminates in a sharp prickle. Cones appear singly, in clusters, or sometimes even in whorls. Trees of some provenances begin to produce cones at a very early age (six years). The tree is a pioneer species, a short-lived colonizer of bare ground. The subspecies *latifolia* is particularly well suited to reforestation of fire sites; the heat of a fire, while destroying the parent tree, will initiate cone opening and seed distribution on to a bare site shortly after the fire is out.

Growth conditions

The value of this tree in British forestry is its hardy constitution and low nutrient requirements. Few, if any, other forest trees will tolerate such poor, acidic, exposed upland sites in northern and western Britain. Even on such sites, the top growth may be so good that the tree becomes unbalanced, especially with coastal provenances. The heavy tops of 5–10-year-old trees cannot be properly supported by their relatively small root systems on soft wet ground, and the plants begin to lean over at an angle, growing vertically again in subsequent years. The resulting 'basal sweep' in the trunk is permanent and makes management and harvesting difficult. Inland lodgepole pines do not usually grow strongly enough to be affected. There is no advantage in bringing this hardy tree down to lowland sites; although it does tolerate alkaline soil in parts of its natural range, it will not compete with higher production or quality forest trees in Britain. In very hot weather the small light seeds are carried some way from their parents by very efficient aerodynamic wings, and natural regeneration can occur over a wide area if the ground is suitable.

▲
Cones have sharp little spines on them which makes them difficult to pick.

Propagation and management

Lodgepole pine plants are usually raised in open nurseries, preferably on light, acid, sandy sites. Seeds are small in size and weight, and some 300 000 can be expected per kilogram, producing about 120 000 usable seedlings in one year. Cones must be picked in October and November. The light-demanding nature of the tree ensures that lower branches in a dense plantation soon die off, leaving small knots. If spaced out too early, however, coastal provenances in particular will sometimes retain thick heavy branches and rough stems. Conversely, if they are not thinned in time, production of wood will diminish and the weaker individuals in the crop may die out. Growth rates vary, but are never outstanding. The tree's longevity in Britain has not been fully tested because the introduction date is so recent; pioneer species are, however, not generally long-lived. Even pampered specimens in collections and gardens tend to break up before reaching the age of 100 years. Various pests, particularly insects, are often a problem in British lodgepole pine monoculture.

Timber quality and uses

In Canada, stands of the inland form of the tree are now extensively logged, but little timber is exported from North America, and it does not appear to have a trade name other than its own. The timber is light, uniform in colour, slightly resinous, and has fine grain. The heartwood is reddish or brown, and the sapwood is pale yellow. In North America exposed sand dunes are sometimes fixed with this tree, where it resists salt spray and violent winds, but dunes with a very high alkaline sea-shell content cannot be successfully stabilized with it. Timber produced in Britain has so far been of low value, with frequent bends and large knots concentrated at the nodes.

▲
The timber is strong, resinous, and can often be knotty.

27

Scots pine

Pinus sylvestris

Scots pine

Pinus sylvestris L.

Pinus is from the Latin and refers to the pines, and *sylvestris* means 'woodland'. 'Scots' refers to one of the natural European locations.

²/₃ actual size

Origin and distribution

The wide natural range of the Scots pine is broadly similar to that of the silver birch, with which it is closely associated in nature. The range extends across Europe from the British Isles to Spain, Asia Minor, and across Russia to the Pacific coast. Populations of trees on the fringes of this huge area have generally been given variety status because of slight dissimilarities from the type. Native Scottish trees, for instance, belong to the variety *scotica* (Willd.) Schott, and have finer bark, symmetrical cones, and very blue needles. In England, Wales, and parts of Scotland, a considerable amount of Scots pine is of introduced or naturalized European stock. The largest and best specimens in Britain are to be found in Scotland, the tallest being 36 metres (118 feet) and the largest trunk diameter 169 centimetres (17 feet 5 inches in girth) at breast height in 1987. Strongholds of an ancient, but much depleted, native 'Caledonian' Scots pine forest can be seen in Glen Affric, Glen More, and Rannoch. For some years now, all new pine trees planted in and adjacent to these areas have been from authentic local stock to preserve the genetic resource.

▲
Isolated old Scots pines on heathland indicate past heathland fires and animal grazing.

Key features

The Scots pine commonly reaches heights of 25–30 metres (nearly 100 feet) with fairly straight cylindrical stems. Few stems, even with great age, exceed 100 centimetres (3 feet 3 inches) in diameter. In early life, the outline of the tree is conical with whorled branches, tending to be upswept towards the extremities. With age a more flat, irregular crown develops, often on a long, clean, branch-free trunk. Open grown trees, however, may occasionally retain low heavy branches and green foliage to the ground. The young shining bark is distinctly reddish or orange in colour, papery and tending to peel, by the age of 20 years becoming darker brown and rugged. Very old trunks often develop a regular pattern of pale pinkish-grey, hard, flat plates clearly defined by dark fissures. The young shoots are greenish, smooth and shining, becoming greyish-brown and prominently marked by the retained bases of scale leaves. The buds are oblong, ovoid, shortly pointed, dull reddish brown, and the bud scales are often free of resin. Needles occur in pairs and last for about four years on healthy trees. They are glaucous green, 5–10 centimetres (2–4 inches) long with minute forward pointing teeth and a translucent pointed tip. A basal sheath around each pair of needles shortens slightly with age and changes

The distinctive ridged bark of a mature Scots pine.

Male flowers are numerous and very obvious in spring.

The pretty little pink female flowers are not easy to find.

from light to dark grey. Male flowers consist of clusters of golden anthers which are pinkish prior to shedding pollen, arranged in groups on a central length of occasional current side shoots. Female flowers appear as tiny crimson globes at the tips of main shoots in spring, particularly towards the top of the tree, which develop into pea-sized conelets in one season, and into full-sized green and then grey-brown cones the following year. Cones may be carried singly or in clusters, each one being 5 centimetres (2 inches) long, a woody and ovoid–conic structure with a short stalk. The cone scales are narrowly oblong with only a minute spine, or its remains, at the end. A pair of seeds develops on each scale. In dry weather, usually in February the cones open and the winged seeds are distributed by the wind. A unique feature of pine seed is that it is attached to the wing by a claw-like structure.

▲
The cones shed seed from January onwards, and fall the following summer.

Growth conditions

The Scots pine, with its broad natural range, has adapted to many different environments. Its range extends beyond the Arctic Circle in the north (to 71°N) and to an elevation of 2600 metres in the Caucasus. It thrives in both humid and arid climates and will tolerate a wide range of soils, even stabilized sand dunes. It is totally hardy in Britain and, unless badly managed, it is wind-firm. Although the Scots pine grows naturally in close proximity with its fellows, it does require a fair amount of light and is soon suppressed without it. This tree is typically a dry sandy heathland species, but it grows very slowly on impoverished sites.

Propagation and management

Propagation is normally by seed which is rarely imported into Britain. Trees usually produce a crop of cones from about the age of 16 years and maximum cone production occurs between 60 and 100 years. There is normally an interval of 2–3 years between heavy cone crops. If seed is to be collected, the green cones must be picked in January before they open naturally. There are some 165 000 seeds per kilogram, about 140 000 of which should be viable, and from which 63 000 seedlings can be expected after one year. The planting season for bare-rooted plants is a long one because the shoots do not extend until the beginning of May. Stand management must be good if disease and insect problems are to be avoided, and stump treatment is required after felling in plantations to inhibit the spread of fungal root disease. This involves painting freshly cut stumps with spores of a wood-rotting fungus to make them less palatable to more destructive fungi, which may kill adjacent live trees. Growth rates of the Scots pine do not compare favourably with those of the fast-growing spruces or the Corsican pine, but a leading shoot of 30–40 centimetres (12–16 inches) can usually be expected to grow each year from thicket stage to pole stage. Trees retained into old age may produce very-high-quality, straight, knot-free timber. The most productive plantations will need to be thinned out from about the age of 22 years and are likely to have an economic rotation of about 70 years.

Timber quality and uses

Commercial names for the timber of Scots pine include red deal, yellow deal, baltic fir, or confusingly just 'redwood'. There are some 17 trade names in use, often recalling the port of shipment. The wood is pale reddish-brown with distinct rings and a moderate amount of resin. The sapwood is pale coloured and forms an uneven band around the heartwood. Although not durable without treatment, the timber is strong, light and is easily wrought. Nor is it difficult to season, and it is stable enough to be used extensively for joinery and constructional work. In the past, huge quantities of Scots pine timber were used for railway sleepers, transmission poles, and pitprops. In recent years there has been an increased demand for pine furniture, where knots are thought of as a decorative advantage rather than a potential weakness.

Most pine plantations in Britain are composed of one of only three species. In addition to the Scots pine, which is distributed throughout Britain except on soils with free chalk, there is the Corsican pine usually seen in the southern half of the country and the lodgepole pine, which prefers the acidic soils of the northern hills. All three are in the group which carry their needles in pairs. Occasionally small woods or plantations of more exotic species are planted. In mild locations the bishop pine, *Pinus muricata*, which also has its needles in pairs, and the Monterey pine, *P. radiata*, with needles in threes, have been planted for some years. On hill sites, the Macedonian pine, *P. peuce* is being grown on an experimental scale, its needles occurring in fives. The western yellow pine, *P. ponderosa*, is a possible substitute for the Corsican pine in some areas.

▲
The timber, known as 'red deal', is commonly used in construction works.

28

Western balsam poplar

Populus trichocarpa

TORREY AND GREY EX HOOKER

Western balsam poplar

Populus trichocarpa TORREY AND GREY EX HOOKER

Populus is from *arbor populi*, the 'tree of the people'. *Trichocarpa* is from *tri* meaning 'three' in both Greek and Latin, and *karpos* is the Greek for 'fruit' or 'carpels'. The common American name is cottonwood because of the cotton-like seed fluff. The name balsam is loosely used for poplars that have balsamic sticky buds.

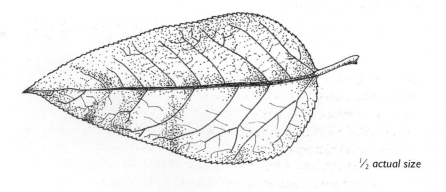

$^{1}/_{2}$ *actual size*

Origin and distribution

The western balsam poplar is a tall graceful tree that grows rapidly on suitable sites in Britain. In North America it has a huge natural range and occurs in many forms. In the north the range extends through western Canada and into Alaska; the southern limits are the coast and islands of California; it extends eastward to the plains of Idaho and Montana, and will even survive up to an elevation of 1800 metres (5900 feet) on the west side of the Sierra Nevada. This tree is frequently found growing with forest conifers such as the Douglas fir on the hills, and American aspen and red alder on the plains. Pure stands occur on unstable river gravels. A record height of 68.5 metres (225 feet) has been measured in America, making this the tallest broadleaved tree to be found there. The western balsam poplar was first introduced to Britain from Canada in 1892. Since then, selected clones have been imported to Britain at intervals, notably from Washington State, to provide vigorous and disease-resistant material for new plantations and for tree breeding. The tallest specimen on record in Britain has reached 41 metres (134 feet), and the largest stem diameter in 1987 was 106 centimetres (10 feet 11 inches in girth).

▲
These cottonwoods have reached maturity in about 35 years.

Key features

The crown of the western balsam poplar is rather narrow and conical with ascending branches. The whole tree, although large, has a fairly light and open appearance because of the pale grey-brown, often silvery, bark and whitish-green-backed leaves. Older trees have slightly fissured stems often with frequent bunches of epicormic shoots. The lifespan of this species is relatively short—plantations reaching 30 years are considered old. On good sites and using good clonal material, the maximum height can be reached in half that time, after which the branches begin to deteriorate and the timber ring width is much reduced. The buds are viscid and strongly balsamic and are carried on angular glabrous, ochre-grey shoots. The bud scales are shining brown and not tightly packed together. The leaves are ovate and pointed, dark

▲
Bright red, male catkins appear in March.

▲
The silvery backed leaves distinguish the balsam poplar.

glossy green above and conspicuously silvery-green below, and sometimes rust-coloured near to the prominent pale veins. The foliage starts off butter yellow in early April and in autumn turns to old gold in colour but still showing white-backed leaves. Male trees bear thick, dull, red catkins before the leaves unfold. Female flowers, on separate trees, are green and evenly spaced out along 15-centimetre (6-inch) catkins. In May and June the minute seeds packed in loose 'cotton wool' fly away on the wind in vast quantities. Although seed production is massive, survival is often negligible because ground and climatic conditions for germination are critical and the period of viability is very short. To compensate, the tree also reproduces itself vegetatively in nature by shedding branches which root after falling into shallow water or soft wet earth.

Growth conditions

Clones of the western balsam poplar imported currently for use in Britain are hardy and disease resistant, but they do require rich soils, considerable space, and a lot of moisture. This species and its hybrids tend to be more tolerant of acidity than the black poplars, but will not tolerate acid moorlands. The rapid top growth quickly suppresses competing vegetation, but the narrow crowns of widely spaced trees will allow species such as elder, bramble, and common alder to form a shrub layer of sorts, providing food and protection for some wildlife. If planted closely together, this species will totally dominate the site in much the same way that an agricultural crop would do.

Propagation and management

Propagation of the western balsam poplar in Britain is always by cuttings and only tested and registered material should be used. It has been suggested that only 5 per cent of all *trichocarpa* are naturally resistant to the bacterial canker *Xanthomonas populi*. Existing plantations of

balsam poplars, particularly some of the hybrids, can still be seen with the severe symptoms of this canker, which are swellings and openings on the twigs and branches from which the whitish (and then black) bacterial slime exudes. Over several years, large rough swellings occur and branch ends die and fall off. Rust fungi such as *Marssonina* and *Melampsora* can also be a problem, causing premature defoliation which may result in reduced resistance to autumn and winter frosts. In the absence of rust fungi, frost damage to poplars usually does not spoil the foliage, but it can still cause longitudinal stress cracks in the main stem. These occur mostly on the south side of the tree where sudden warming takes place after a cold night.

▲

The fertile valleys in the Forest of Dean can make excellent poplar woodlands.

This coloured timber is attractive, but clean white wood is more in demand.

Timber quality and uses

This poplar had many traditional uses in America similar to those of the American black poplar *P. deltoides*: their timber is soft, impact absorbing, and fire resistant. The western balsam poplar in Britain today is used both for timber and for biomass production, that is a short rotation coppice crop which is chipped for fuelwood. It has already played an important part in tree breeding research and hybridization with several other species (notably *P. deltoides*), and there are a number of new selections being researched as potential high production trees. It is not suitable for ornamental use or urban planting, and should never be planted near to buildings.

29

Hybrid black poplar

Populus × canadensis MOENCH

(*P. × euramericana* (DODE) GUINIER)

Hybrid black poplar

Populus × canadensis MOENCH
(*P. × euramericana* (DODE) GUINIER)

The genus name *Populus* is derived from *arbor populi*, the 'tree of the people'. *Populus* is also the Latin form of the Greek *papaillo* meaning 'to shake,' a reference to fluttering leaves. The hybrid and species names used here are *nigra* meaning 'black', *deltoides* which in Latin is triangular and refers to the shape of the leaves, 'serotina' meaning 'late', referring to the lateness of leafing, and 'robusta' which is a Latinized word for 'robust'.

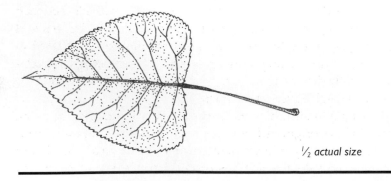

½ actual size

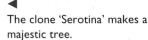

The clone 'Serotina' makes a
majestic tree.

Origin and distribution

A large group of hybrids have been produced between the American black poplar or cotton-wood *Populus deltoides* and selected clones of varieties of the European black poplar, *P. nigra*. These hybrid trees, which form the majority of small poplar plantations and screens in Britain, are also called *Populus × euramericana*. Although not strictly botanically correct this name is used by the European Poplar Commission to cover all the black poplar hybrids between American and European species, including the Carolina poplar, *P. angulata*. In order to avoid confusion and implied geographical origin, the International Poplar Commission decided in 1950 to use the collective name × *euramericana*. Using more up-to-date horticultural nomenclature still, both species epithets are usually dropped and named clones are referred to simply as *Populus* followed by the clone name in single inverted commas.

Almost all hybrid black poplars belong to quite distinct clones which are identified by name or a number. The first clone known to have originated from the *nigra × deltoides* cross was culti-vated in France in 1750. In Britain the most widely planted of these clones is probably 'Robusta', a male selection with a straight stem which has reached to a height of 37 metres

(121 feet) in Devon. Another common tree in Britain and the largest poplar recorded here so far is 'Serotina', the Black Italian poplar which is itself further divided into clonal selections. 'Serotina Aurea', for instance, is the well-known ornamental golden poplar, common in town parks and large gardens. The largest ordinary 'Serotina' on record in 1988 was 175 centimetres diameter (18 feet in girth) growing near Chepstow. Some of the tallest trees have reached 46 metres (151 feet) in southern England. The female clone 'Regenerata' was widely planted around 1814 and is the result of a back cross between 'Serotina' and *P. nigra*. Its distinguishing feature is that new leaves are green, and not the olive brown colour common to most of the other poplars. It also produces huge quantities of seed fluff which is often a nuisance.

▲
Female flowers are yellow in March and male flowers are usually deep red.

Foliage varies from clone to clone, but the leaves are broadly triangular.
▼

Key features

In all its forms, *Populus × canadensis* is a tall deciduous tree with more or less ascending branches, at least until middle age. Older trees may become massive with heavy, sometimes arching, limbs and an angular outline. The bark is variable but generally rough and somewhat oak-like; in old age deep vertical fissures develop. The twigs are usually angular in their first year, smooth and pale yellowish grey-brown in colour. They are glabrous, or only slightly pubescent at first on some clones. The leaves are deltoid, almost triangular and truncate at the base, and there are usually one or two glands near the leaf stalk. Through the summer the green or grey-green foliage flutters in the slightest breeze, and in a strong wind becomes very mobile, making a sound reminiscent of running water. This excessive leaf movement results from a combination of the unstable triangular leaf shape and the long vertically flattened leaf stalk. The buds are slightly sticky and smell faintly of balsam. *P.* 'Eugenei', which was often planted in the 1950s and 1960s, has a narrow pyramidal crown shape only slightly less compact than the familiar Lombardy poplar. *P.* 'Marilandica', which was popular at about the same time, is less ascending and, unlike most of the other clones, is female. Black poplars are generally thought of as short-lived (30–50 years), perhaps because they can be grown on short commercial rotations. When planted as ornamental specimens, black poplars may stand for well over 100 years, but by then the branches tend to become weak and unsafe, and the root spread could be enormous.

Growth conditions

Poplars are demanding trees, requiring full light, ample ground water, and good nutrition. Alluvial valleys are likely to provide the best conditions for the production of high-quality timber. In spite of their fast growth and huge bulk above ground, poplars are wind-firm, and will only blow over in exceptionally violent storms, or where rooting has been restrained. New trees may take a few years after planting to become stable. Acid sites are quite unsuitable for hybrid black poplars and exposure will severely restrict growth. On lush streamside sites, head-high vegetation may be expected between the trees for several years before it is partially shaded out. Poplar leaves soon rot down on the ground, and are then converted to rich loam by the action of worms.

▲

Young, black poplar stands produce good, straight poles in a very short time.

Propagation and management

Propagation is normally achieved by winter cuttings taken from fast-grown, one-year-old shoots. These are most conveniently obtained from special root stocks in 'stool beds', cut each year for this purpose. Cuttings of current wood about 30 centimetres (1 foot) long are inserted vertically into a well-cultivated nursery bed until the top is about level with the surface of the ground. This work is best done in late winter after the most severe ground frosts have passed, but before the end of March. Rooted plants between 1.5 and 2 metres (up to 6 feet 6 inches) may be expected after one year. The poplar is intolerant of any weed competition and young plants are normally mulched or chemically weeded. Trees should be spaced 8 × 8 metres (26 feet) apart, except for pulp production or short rotation coppice, for which the western balsam poplar or its hybrids are usually better suited anyway. Under EU regulations, plants for forest use must be raised in approved and registered locations. For several years after planting out, pruning and removal of epicormic regrowth from the lower stem is essential. Poplar roots on clay soils are particularly known to have a drying and shrinking effect, and sites have to be carefully selected well away from buildings. Moisture and fertility are required, and shelter from wind will reduce post-planting instability.

Timber quality and uses

Poplar once had many traditional uses ranging from wagon bottoms and floorboards to food containers. Veneers of the light-coloured soft wood were once woven into disposable fruit punnets. The wood is clean and does not taint. Although hybrid black poplars are less productive than some balsam poplars, they should reach 30 metres (98 feet) in about 25 years. The British match industry used to consume huge quantities of knot-free poplar wood until 1960, but it has now declined. New markets include vegetable crates and other food packaging products, and medium-density fibreboard, as the timber absorbs impact rather than splitting, making it ideal for these uses. As living trees black poplar are very useful. They will establish themselves very quickly and easily, and produce an almost instant screen of trees. Given very careful management, this screen can subsequently be used to nurse longer-lived and more valuable species. Poplars are moderately resistant to air pollution, but few towns have room to accommodate their wide-ranging root systems and huge bulk. A large 'Serotina' may contain 36 cubic metres (1200 cubic feet) of timber.

▲
The timber is usually clean, straight, and pale-coloured.

30

Redwoods

Sequoia sempervirens (D. DON.) ENDLICHER (COAST REDWOOD)
AND *Sequoiadendron giganteum*
(LINDLEY) BUCHOLZ (WELLINGTONIA)

Coast redwood

Wellingtonia

Redwoods

Sequoia sempervirens (D. DON) ENDLICHER (COAST REDWOOD) AND *Sequoiadendron giganteum*

(LINDLEY) BUCHOLZ (WELLINGTONIA)

The genus name *Sequoia* comes from a Cherokee scholar's name *Sequoyah* (1770–1843). *Sempervirens* is from the Latin *semper*, 'always', and *vivo*, 'alive', alluding to the evergreen nature of the tree. *Sequoiadendron* is a combination of two names: the tree was first assigned, wrongly, to the genus *Sequoia* to which the Greek *dendron*, meaning 'tree', was added to distinguish it from the coast redwood. Wellingtonia commemorates the Duke of Wellington who died at the time the tree was introduced to Britain. The wood of both trees is indeed red.

Coast redwood

Wellingtonia

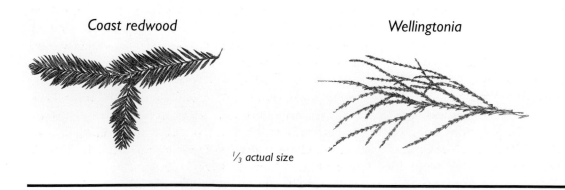

$^1/_3$ *actual size*

▲
Big stems occur in Britain which evoke the atmosphere of the American redwood forests.

Origin and distribution

The redwood genus *Sequoia* only contains one species *sempervirens*. The redwood was probably widespread in preglacial times, but in the last 10 000 years it has been confined to a narrow, mountainous, coastal strip of California from Monterey northwards just into Oregon. It occupies a 'fog belt' where summer fogs off the Pacific Ocean enable it to grow in an otherwise unsuitably hot dry climate. The best trees are found north of San Francisco, up to an altitude of 910 metres (3000 feet). The redwood was collected by Archibald Menzies in 1796 and first described by David Don in that year, but fifty years passed before it was eventually named by Stephan Endlicher and introduced to Britain by Theodore Hartweg. The tallest tree of any species ever accurately recorded in the world is a redwood, having reached 112.4 metres (368 feet 8 inches) in 1977. The tallest *Sequoia* in the British Isles was 47 metres (154 feet) at Bodnant in 1984, and the stoutest, 245 centimetres diameter at breast height (25 feet 3 inches in girth), was measured in 1989. *Sequoiadendron giganteum* also originates from a limited range in California, on the western side of the Sierra Nevada at altitudes of 1400–2100 metres (4500–7000 feet). Isolated redwood stands are separated from each other by former glaciation sites which, since the retreat of the ice sheet, have not been colonized by redwood trees. There were two separate introductions of *Sequoiadendron* to Britain in 1853. John Matthew introduced it to Perthshire and William Lobb brought some seeds to the nursery of Messrs Veitch in Chelsea, but a much larger import by William Murray in 1859 accounts for most of the big redwoods that we see today in Britain. The spread of *Sequoiadendron giganteum* here as an ornamental tree has been quite extraordinary, and it can be found at almost every stately home throughout the length and breadth of the country. Redwood avenues are commonplace, and clumps or roundels are often seen. Woods are less common, but they are often quite spectacular, with huge trees giving the impression of vigour and rapid growth. The tallest tree known in 1991 was 53 metres (174 feet) tall at Castle Leod, and the largest stem diameter at breast height in 1988 was 345 centimetres (35 feet 6 inches in girth) at Clunie in Tayside. In America, the largest recorded sizes are 85 metres (279 feet) tall and 800 centimetres diameter (82 feet 5 inches in

girth). The oldest specimens there are reckoned to be about 3500 years old. Few trees appear to die naturally, so we can assume that ultimate age and size has still to be reached. In Britain, diameter growth continues at an impressive pace but, in the south particularly, vertical growth tends to slow down or even stop after about 120 years.

Key features

The most obvious feature of *Sequoiadendron* (Wellingtonia) is the massive stem, supported at the base by curving buttresses and covered with soft, brownish-red, spongy, fibrous bark. Isolated trees may have branches reaching to the ground where, given the opportunity, they will take root and send up a ring of subsidiary stems. In some circumstances, a small wood may be created from one tree. Coast redwoods *Sequoia sempervirens* will do the same from coppice regrowth. In close proximity, low branches soon die off, leaving columns which are impressive, clean, red-barked, and less buttressed than Wellingtonia, reaching high into the ever-green canopy. The coast redwood needles vaguely resemble those of the yew except that they are shorter and lighter green in colour. Wellingtonia foliage is thread-like with numerous, tiny, overlapping scales.

▲
Most Wellingtonia trees are in gardens and arboreta like this 1860 specimen at Westonbirt Arboretum.

Male flowers of both redwoods are small and numerous, carried on side shoot tips, and shed their pollen in February. Female flowers are yellowish-brown spiny globes, borne singly or in small bunches on short shoots. These develop into oval, green, woody cones which hang on the trees for several years. Numerous flattened, pale brown, shining seeds are produced between the blunt-ended cone scales. Due to the cold February weather when the pollen is shed, seed in Britain is usually infertile.

Growth conditions

The Wellingtonia grows well over most of Britain, where it survives rather better than the coast redwood. It is not a mountain tree in this country and is never seen in such areas nor is it really a town tree. Although they do survive, many of the specimens seen in towns are quite appalling because it hates compacted ground and turbulent winds around city buildings, and is sensitive to air pollution. Trees that do survive in urban situations are often stunted and unhealthy, quite unlike the same species growing out in a cool, damp, clean forest. The redwood is hardly affected by exposure to wind and seems almost immune to wind-throw, although in strong winds it will shower down brittle dead branches and bits of foliage. It tolerates most soils except for acid peats and sandy heathland. It requires full light at every stage of growth, and casts a heavy shade, although nettles, brambles, and elder bushes commonly survive under it.

▲
Coast redwood needles are flat and fairly short.

Relentless upward growth usually takes this tree beyond the height of anything else around it, so it is prone to lightning strikes. They usually result only in the temporary loss of the top branches, but on occasions whole trees may be torn apart, with devastating effects. The coast redwood is a tree more suitable for moist loams in western Britain.

Sequoiadendron forest is unusual in Britain; these trees are at Lynford in Norfolk.
▼

Propagation and management

Redwood seeds are very small, with little in the way of resources, and rapidly dry out or rot away. Seed collected in Britain is seldom of any value, natural regeneration is hardly ever seen, imported seed is variable, and germination is uncertain. There are a huge number in a kilogram, and it is impossible to estimate how many or the likely production even at the end of one year. In America seeds may remain alive in the closed cones for up to 20 years. In nature these species are both dependent upon forest fires, partly to clear the forest floor as a seedbed and partly for the heat needed to open the cones and shed the seed. Parent trees are insulated from the fires by their thick bark. The coast redwood can also be propagated successfully by cuttings, but these must come from upward-pointing foliage. Cuttings must be protected from strong sunlight and early or late frosts, and competing vegetation in the forest must also be controlled in the early stages of growth.

Timber quality and uses

The heartwood colour is deep pinkish-brown with very distinct growth rings of contrasting colours and density representing early and late wood. It is straight grained, nonresinous, and light. Knots are almost nonexistent on old stems, but timber from the core of the tree or near the top may contain very large knots indeed. Redwood timber has a low bending strength, little stiffness, and does not resist shock. It dries rapidly and, once dry, remains fairly stable. It is not durable and must be treated with preservative if used outside or in contact with moisture. Stain, although absorbed freely, gives a good finish.

A close relative of the redwoods is the Japanese cedar *Cryptomeria japonica* (L.fil.) D.Don, which is a striking tree with vertically shredding, red, hard bark and evergreen cord-like foliage. It was introduced to Britain in 1842 from China and in 1861 from Japan. As with the redwoods, pollen is shed in February and, because of the harsh weather in Britain, is of little use here. The tree requires wet sites but is unstable on soft ground. It will blow down, bend over, or even snap very easily if planted in an exposed place. Cut stumps coppice freely. Plantations of *Cryptomeria* are of little value to wildlife or nature conservation, and there are no recognized uses for the timber in Britain. In Japan it has the commercial name *Sugi* and is one of the most important softwoods produced there. It is completely stable and is used for buildings and structural work, for joinery and cooperage, and even boat building.

▲
The beautifully coloured Wellingtonia timber has, in fact, little value.

31

Norway spruce

Picea abies

(L.) KARST.

Norway spruce

Picea abies (L.) KARST.

Picea is from the Latin *pix*, referring to the pitch or resin produced by some spruces. *Abies* harks back to a time when this tree was assigned to the *Abies* or fir genus. The common name describes a spruce tree widely grown in Norway. Spruce is from *pruce*, an old name for Prussia.

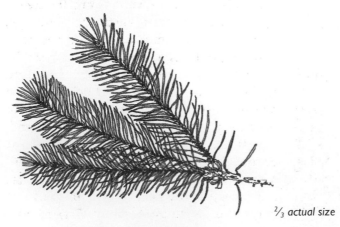

²/₃ *actual size*

Origin and distribution

This common and widespread tree has a broad natural range covering central and northern Europe, Scandinavia, and north-west Russia. In Poland, Romania, and south-eastern France the Norway spruce is confined to the mountain ranges. It is not a native species in Britain; its date of introduction is uncertain, but a reference was made to it in the mid-sixteenth century by Turner in his *Name of Herbs* published in 1548. It is widely naturalized and seeds freely on the edges of plantations and forest roadsides. In such situations the shallowly rooted seedlings suffer greatly from drought in the first year or two. Planting has been widespread in the British Isles, but numbers have always been less than the Sitka spruce. In areas suitable for growing spruce, the two species would often be placed side by side. Norway spruce would traditionally be planted on the low valley sides, wet flushes below spring lines, and generally more fertile ground, while the Sitka spruce would be planted higher up into the hills. Care was still needed, however to avoid frost hollows. The largest trees known in Britain are found in Cumbria and Scotland. The greatest height recorded in 1986 was 52 metres (170 feet) and the largest diameter at breast height is 147 centimetres (15 feet 2 inches in girth). In general, spruce trees seldom reach anything like these sizes but, on a good site, heights in the region of 20 metres (66 feet) may be expected in only 35 years.

Key features

The Norway spruce is usually conical in shape and has rich deep-green needles. The branches reach down to the ground when the tree is grown in isolation but, as a component of a forest, the green part of the crown is small and confined to the pointed top. A tracery of persistent, fine, tough, dead branches remain on the stem all the way down for many years. The bark is smooth, and thinly scaling, reddish- or orange-brown, becoming grey only in old age. From some distance away, Norway spruce forests below the green canopy have a distinctive rusty-

The flowers are seldom seen; the females congregate at the tops of the trees.
▼

red look; perhaps because of this the tree is called 'red spruce' in Germany. With age, large butresses form at the base of the stems, and long roots often develop on or near the surface of the ground. Current shoots are dull orange-brown above and golden brown beneath, usually glabrous. The whole shoot is sharply ridged and closely covered with woody leaf pegs. The buds are ovoid, shiny, or purplish and covered with translucent resin. The 2-centimetre (¾ inch) long needles are glossy mid-green, not at all glaucous like those of the Sitka spruce. The needles radiate from most shoots but become pectinate on older or suppressed branches.

▲
This young cone is at the pretty purple stage in midsummer. By autumn it will ripen to dull brown.

Healthy needles are rigid, quadrangular, and pointed; they are retained on the tree for three, four, or five years if not attacked by spruce aphids. Mature branches may develop curtains of pendulous side shoots weighed down by thousands of needles. The flowers, in common with most conifers, are simply clusters of male stamens shedding pollen in May, and the pinkish- or pale-green cone-like females are confined to the very tops of the trees. Female flowers always mature and have finished before the males on the same tree ripen, which ensures cross-pollination. The ripe cones are pendulous red-brown cylinders about 15 centimetres (6 inches long), with leathery scales which have smooth edges and are hardly reflexed. Each cone is blunt-ended and jagged as if it had been bitten off. The winged seeds are held in a cup-like basal extension to the membranous seed wing. The Norway spruce does not appear to be a long-lived tree, and almost all of those known to have been planted before 1850 have disappeared. Most of those that escape the commercial forester's chainsaw eventually fall victim to heartrot in the bottom metre or so of the trunk, which invariably causes the trunk to shear off at that point and fall over.

Growth conditions

The Norway spruce is not suited to very chalky or poor acid soils, but it is more tolerant of lime than most conifers. It cannot tolerate a dry climate unless there is abundant water in the soil. New growth begins on different trees over a period of about three weeks and, although late flushing individuals are not damaged by spring frosts, a spruce forest will not thrive in particularly nasty frost hollows. The Norway spruce will not tolerate very much exposure to salt-laden winds or salinity below ground. It is, however, very much at home on wet hillsides, clay soil, and the more fertile peats. It is well suited to ground that may otherwise be considered suitable for the Japanese larch and Douglas fir, and may replace the Sitka spruce on some inland sites. Except for spring frost damage, the Norway spruce is totally hardy in the British Isles. There is a possibility that it may blow down on soft ground after reaching pole stage, particularly if early thinning and drainage are not very carefully managed. On some sites thinning precipitates a condition leading to crown dieback. Plantation trees with confined space may not stand up either to gale force winds or the weight of clinging snow. Forests of Norway spruce provide dense year-round cover for many small birds and animals. Goldcrests and long-tailed tits find shelter and food in the treetops. During the brief thicket stage, young lowland plantations are often a rich habitat for wild flowers, nesting birds, mammals, reptiles, and amphibians.

▲

Shade-tolerant Norway spruce stand close together in forest conditions.

Propagation and management

The Norway spruce will regenerate naturally but the survival of such progeny is variable. Nursery stock must have a well-balanced, fibrous root system, or losses from drought in the first year may be high. Shallow planting is essential because, throughout the life of the tree, most of the root system never goes far below the ground. Seed is usually collected just before the cones open in October. Fertile seed is not produced by trees less than about 30 years old. There are some 145 000 seeds to the kilogram, from which 55 000 usable seedlings may be expected in one year. Seed from numerous provenances across the huge natural range of the species have been tested. Superficially, the differences between them seem to be small, but the rates of growth and time of flushing, and consequently susceptibility to frost damage, may be dramatic. Young plants will tolerate a remarkable amount of shade from ground vegetation, but may be spoiled by abrasion or suppression if they have to compete with woody weeds for a long period. The main enemies of Norway spruce forests are aphids, beetles, and fire, but protection may also be needed from rabbits, deer, and even theft for Christmas trees. Timber crops are usually harvested on a short rotation of around 30–40 years as poles for pulp, or a medium rotation for sawn structural timber.

Timber quality and uses

Two of the trade names for this leading world timber are 'Whitewood' and 'White Deal'. It is used in many kinds of box and packing-case manufacture, building, and joinery, and has a natural sheen or lustre, particularly when quarter sawn. It is a clean, fairly resin-free, softwood, well suited to food packaging, chipboard, wood wool manufacture, and paper pulp. Provided that it is treated with preservative (which it takes very well under pressure), it is useful for light agricultural and estate work out of doors. Young, fast-grown, sawn timber may tend to

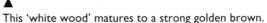

This 'white wood' matures to a strong golden brown.

tear, but otherwise it is easy to work and takes a good finish. The best knot-free pieces may be selected for violin fronts. The tops of thinnings and young trees, rooted or not, are sold for Christmas trees. In most circumstances, the Norway spruce has no special merit as an amenity tree, and felled or partly felled plantations are particularly difficult to disguise in the landscape. These sites usually remain quite bare of any ground vegetation for two years or so.

In Britain there are also trial plots and many isolated park and garden specimens of the Serbian spruce, *Picea omorika*, which is a valuable tree similar to Norway spruce that is likely to be used more frequently in the future. It has a limited and now much reduced natural distribution in the Dinaric Alps of the former Yugoslavia and in the wild it has become an endangered species. It is a botanical anomaly insofar as the other flat-needled spruces are found thousands of miles away in countries near the Pacific Ocean, both in the Far East and in western America. Its outstanding feature is its strikingly narrow outline, which has evolved to shed snow. The branches sweep downwards from the main trunk and then curve gently upwards towards the tips. The sprays of branchlets also have a pendulous tendency. The foliage is dark glossy green and the shoots are dull pinkish-brown with some dark brown pubescence, unlike the Norway spruce. The flat needles are numerous but solitary, tending to curve and having two broad whitish bands on their undersides.

In Britain the Serbian spruce has been tested on a wide range of site types and appears to grow reasonably well on most of them. It will tolerate competition from dense *Calluna vulgaris* to a slightly greater degree than the Sitka spruce, and can survive further into frost hollows than the Norway spruce.

The natural habitat of the Serbian spruce is high limestone mountains, so it is not only hardy but also lime-tolerant. It does require a moist environment for good development, but is probably less demanding in this respect than the more common spruces planted in Britain. It can be used as a Christmas tree or for farm or estate use, it can provide light posts and rails if treated with preservative.

32

Sitka spruce

Picea sitchensis

CARR.

Sitka spruce

Picea sitchensis CARR.

Picea is from the Latin *pix* referring to the great amount of pitch or resin produced by some species. 'Sitka' and *sitchensis* refer to the small seaport in Alaska from which the tree takes its name.

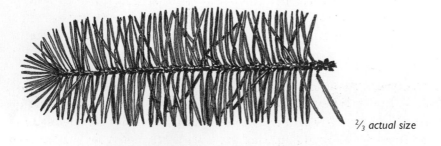

²/₃ *actual size*

Origin and distribution

In its natural state, the Sitka spruce's range extends 3000 kilometres along the west coast of North America, from Kodiak Island in Alaska southwards to California. It was discovered by Archibald Menzies in 1792 at Puget Sound in Washington State, and was eventually brought into cultivation in Europe by David Douglas in 1831. The Sitka spruce has adapted to a vast range of climatic conditions within its long north–south range, from latitude 61°N near Valdez in Alaska to 39°N in California. In the south it is confined to the coastal 'fog belt', and its elevation range is from sea level to 900 metres (2950 feet). In Britain, provenances from this huge area can be selected to suit any western or northern site. Planting is only practicable, however, where moisture, either as rainfall, persistent mist or fog, or in the soil, is adequate. Many of these requirements are met in those regions of western Britain where most 20th century forestry has been practised. The Sitka spruce is planted mostly on cold wet grassland and peat moors; many of these sites have been impoverished by prolonged overgrazing or burning, and often altered by mining, erosion, peat cutting, or lack of drainage. Few other trees would tolerate such places. Even so, on the poorest sites, the lack of nitrogen or phosphorus in the soil may need to be augmented by applications of

Growth in high rainfall areas is rapid; this stem is 25 years old.

artificial fertilizer. Newly planted trees will not compete with a dense growth of heather *Calluna vulgaris*. The largest trees recorded are in Scotland; the tallest in 1990 was 61 metres (200 feet) andthe largest stem diameter at breast height is 257 centimetres (26 feet 6 inches in girth).

Key features

On favourable sites this most vigorous tree will grow upwards well in excess of 1 metre (3 feet 3 inches) per year. Beyond 30 metres of height, upward progress is very much slower. Exposed trees are untidy pyramids of foliage with rapidly tapering trunks. Forest-grown trees have

▲
The cones are papery, and resemble crinkled parchment.

▲
Sitka spruce flowers are hardly ever seen; they are mostly confined to the tree tops.

short narrow crowns and lofty straight stems. The whorled branches, although soon suppressed and killed below the evergreen canopy, are tough and persistent. Old trunks are often heavily buttressed, sometimes even bottle-shaped towards the base. The bark is dark purplish grey-brown with smallish, curled, hard scales which become rougher with age. Current shoots are pale buff and glabrous, while the older twigs have numerous sideshoots, and are darker and scratchy with persistent leaf pegs (pulvini). The buds are small, ovoid or rounded, brown, with some resin which may be translucent or purplish in colour. The needles are rigid and sharply pointed, flattened but with a distinct keel on the underside. They are shining green above and have two broad white bands of stomata below. Well-grown foliage has a crisp glaucous look. The flowers ripen in May: the males are globular and the females, which are confined to the tree tops, are like miniature bright pink cones. As these develop they become narrow, pendulous and straw-coloured, some 7 centimetres (2¾ inches) long with thin, crinkly-edged, papery scales. Each tiny seed has a wing about four times its own length.

Growth conditions

The Sitka spruce is essentially a maritime tree, requiring a considerable amount of moisture to thrive. In western and northern Britain it will survive anywhere, from the highest tree line down to sea level, where its tolerance of high soil-water sodium levels is outstanding. It will not tolerate calcium or magnesium in excess, however, and grows poorly on lime-rich soils or chalk. Spring frosts are a problem, particularly with early-flushing provenances planted directly into short grass on old pasture sites. Windthrow may be a problem on exposed, poorly drained sites, or where the roots have been restricted. Stems may snap if a fast-grown stand is suddenly thinned out later than it should have been. As with all extensive areas of monocul-

▲
Young plantations of Sitka spruce are a feature of upland Britain.

ture, Sitka spruce forests may be prone to predation by bark beetles and aphids, which some-times cause serious damage. Fortunately, most of the tree's natural predators have yet to become significant problems in Europe.

Propagation and management

Sitka spruce seeds are very small; once cleaned there are about 400 000 to the kilogram. Germination is usually good, and after one growing season about 160 000 usable seedlings may be expected to survive. Early growth is slow, and seedlings almost certainly have to stand for at least two years before transplanting. Breeders have in recent years developed genetically improved trees which combine high-quality wood with good productivity. The longevity of the species has not been fully put to the test in the British Isles, but the first plants introduced still survive and a few are still growing strongly. Spruce forests, however, tend, to be har-vested on a fairly short rotation because small roundwood is in particular demand, so very few specimens planted recently are ever likely to reach old age.

Timber quality and uses

Sitka spruce timber is light in weight, straight-grained, and fairly strong. It is easily worked, expect for the most knotty lengths, and takes a good, smooth finish, although it is seldom worked beyond rough sawing. The sapwood is pale cream in colour and the heartwood is a little darker, occasionally with a pinkish tint. It was once widely imported from British Columbia under the trade name 'silver spruce', for use in the aircraft industry, and large sawn timbers were also imported for constructional work. The great value of this tree to Britain now is as raw material for

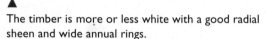
The timber is more or less white with a good radial sheen and wide annual rings.

pulp and paper manufacture, for which the long fibre length of fast-grown wood is ideal. The wood is nearly nonresinous and soft. Although as timber it is not durable, it absorbs preservatives well, particularly under pressure. Seasoning, either naturally or in a kiln, is a quick and easy process. It holds nails well when used for boxes, packing cases, rough joinery, and in building work. Rough logs can easily be reduced to manufacture chipboard and insulation boards. A further refinement of the pulp produces cellulose, and eventually products such as cellophane and rayon.

33

Sycamore

Acer pseudoplatanus

L.

Sycamore

Acer pseudoplatanus L.

The Latin name for *Acer* also means 'sharp' referring to the use of *Acer* (maple) wood for spears, pikes, and lances. The species name *pseudoplatanus,* meaning false plane, indicates that the leaves are like those of the plane tree (*Platanus*). Sycamore is a name borrowed from the eastern fig *Ficus sycomorus.*

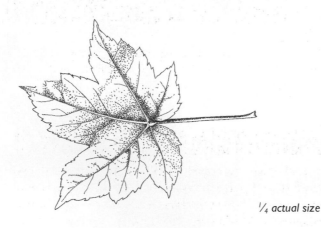

$^1/_4$ actual size

Origin and distribution

The sycamore is truly native only in north-eastern France, southern Germany to the Mediterranean, and eastwards to Asia Minor. It is widely considered to have been introduced to Britain by the Romans or earlier immigrants, but there is no conclusive evidence for this. It was certainly introduced before 1500. In Christ Church, Oxford, there are sycamore leaves carved on the original tomb of St. Frideswide dating back to 1282, but no one knows whether the carver's model was from the continent. Writing in England in the 16th century, John Gerard (1545–1612) considered the sycamore to be a rare tree; however, in Tudor times the Welsh name *Masarnwydd mwyaf* (Great maple) suggested an already long-established tree which had, by that time, been known to reach prodigious proportions in Wales. This, and the commonness of the sycamore in Wales and north-west England, suggests that the earliest introductions were in western Britain. The apparent Celtic origin of the ancient craft of spoon carving from sycamore wood supports this theory. Although not a British native species, the sycamore is now completely naturalized throughout the British Isles. The largest trees recorded in Britain are found in Scotland: the tallest in 1985 was 40 metres (131 feet) and the largest breast height diameter known in 1991 was 232 centimetres (23 feet 11 inches in girth).

▲
This 90-year-old specimen is showing ripples on the bark which could indicate wavy figured wood underneath.

Key features

The sycamore is Europe's largest maple, and trees in excess of 30 metres (98 feet) are common. The bark, even on large trees, is smooth and pale greyish-green, tending to curl and peel off in large hard flakes, exposing clean fresh pinkish-brown areas. The branches are heavy and give the tree a hard outline and solid appearance. The shoots are glabrous, stout and green at first; the large winter buds are also pale green, often with a hint of pink on the outer scales.

▶
The familiar leaves of the sycamore may be up to 18 centimetres across and have five distinctive lobes.

▲
The sycamore flowers in early summer, providing nectar for bees and other insects.

The long-stalked leaves are large, up to 18 centimetres (7 inches) across, usually five-lobed and palmate; the base is cordate (heart-shaped) and the margin coarsely toothed. The colour is dark lustrous green above, but pale and dull beneath. Trees are frequently seen with some degree of purple on the backs of their leaves. The flowers are numerous and hang in large drooping racemes, providing ample nectar for bees. These racemes have female flowers at the base, males in the centre, and sterile florets at the end. The yellowish-green plump fruits occur in winged bunches, most often in pairs, rarely threes. They have hard, membranous wings which are set at an angle of about 60° to each other.

On good sites the sycamore grows very rapidly. For the first 80 years, the girth of the trunk may increase about 4 centimetres (1½ inches) per year, resulting in a tree of about 1 metre (3 feet 3 inches) in diameter.

Growth conditions

The spread of the sycamore over the whole of the British Isles is quite interesting. Although the seeds (borne every year on trees over 35 years old) are numerous, viable and winged, their weight limits the distance that they can travel away from the parent tree. In addition, it is unlikely that seeds taken by animals or birds would remain intact. Thus, it must be concluded that widespread planting has been practised over a very long period. The reasons for this are twofold: first, the tree is exceptionally hardy and is valuable for shelter; second, its timber is extremely useful. Although the sycamore prefers deep, moist, well-drained rich soils, it grows on all but the very poorest land. It is resistant to industrial pollution and exposure to sea spray, and has a long history of shelter-belt planting in northern Britain, extending as far north as Shetland, and to an altitude of 500 metres (1640 feet). In some areas, self-sown sycamores are said to be unpalatable to rabbits which may help to explain success of propagation. The tree has in the past also been used as a coppice species.

Sycamore leaves are often stained with black patches caused by the tar-spot fungus *Rhytisma acerinum*, which is unsightly but fairly harmless. The fungus overwinters in dead leaves lying under the trees.

Propagation and management

Propagation of the sycamore presents no problems at all; in fact, it is usually more difficult to get rid of natural seedlings under a garden tree. There are about 9000 seeds per kilogram, which should yield 2200 usable plants. Although rabbits may not like the young trees, it is still necessary to protect them from damage. Grey squirrels are a constant pest now, particularly in young plantations that have reached pole stage. Main branches and stems are often completely girdled by them in the spring. Sycamores grown for timber can only flourish on fertile, lime-rich soils with some degree of shelter. As an amenity tree it will grow almost anywhere, and is widely used in shelter-belts to reduce the effects of salt-laden winds from the sea.

Sycamore coppice soon develops into a thicket, and later deep, shady woodland.
▼

Timber quality and uses

The timber is very pale cream, almost white. It is strong and hard and can be worked to a very smooth finish. There are numerous traditional uses for it. The earliest in Britain included writing tablets, ox yokes, and ornamental carving. There is evidence of spoon carving from prehistoric times. Celtic design traditions in this ancient craft in Britain are still followed by a few craftsmen. In Scotland, fine boxes for trinkets and snuff were made from sycamore wood, sometimes in conjunction with dark laburnum or imported mahogany. An early use for sycamore was in

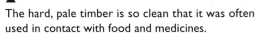

The hard, pale timber is so clean that it was often used in contact with food and medicines.

the manufacture of violin (and related instruments') backs, sides, and stocks. However, since the timber became available in Britain, it has been most widely used where there is contact with food because it is free of dye and does not taint. Kitchen table tops and butchers' blocks, various items in the cider industry, butter pats and prints, and chemists' pestles were made from it. Round turned items such as rolling pins and other machine rollers were often made of sycamore wood. In the textile industry, weavers' shuttles were frequently made from the wood of this tree. Anything requiring a clean, hard-wearing wood that would withstand moisture could be made from sycamore. Later, when supplies could be relied upon, some furniture was made from British sycamore timber.

'Hairwood' is a grey-stained form, and 'weathered sycamore' has been boiled to darken its colour. The timber takes dye well and was used for coloured toys until superseded by beech. More recently the sycamore has been used for plywood and veneer, and there is still a great demand for the wavy grain wood that occasionally occurs.

34 Willow

Salix L.

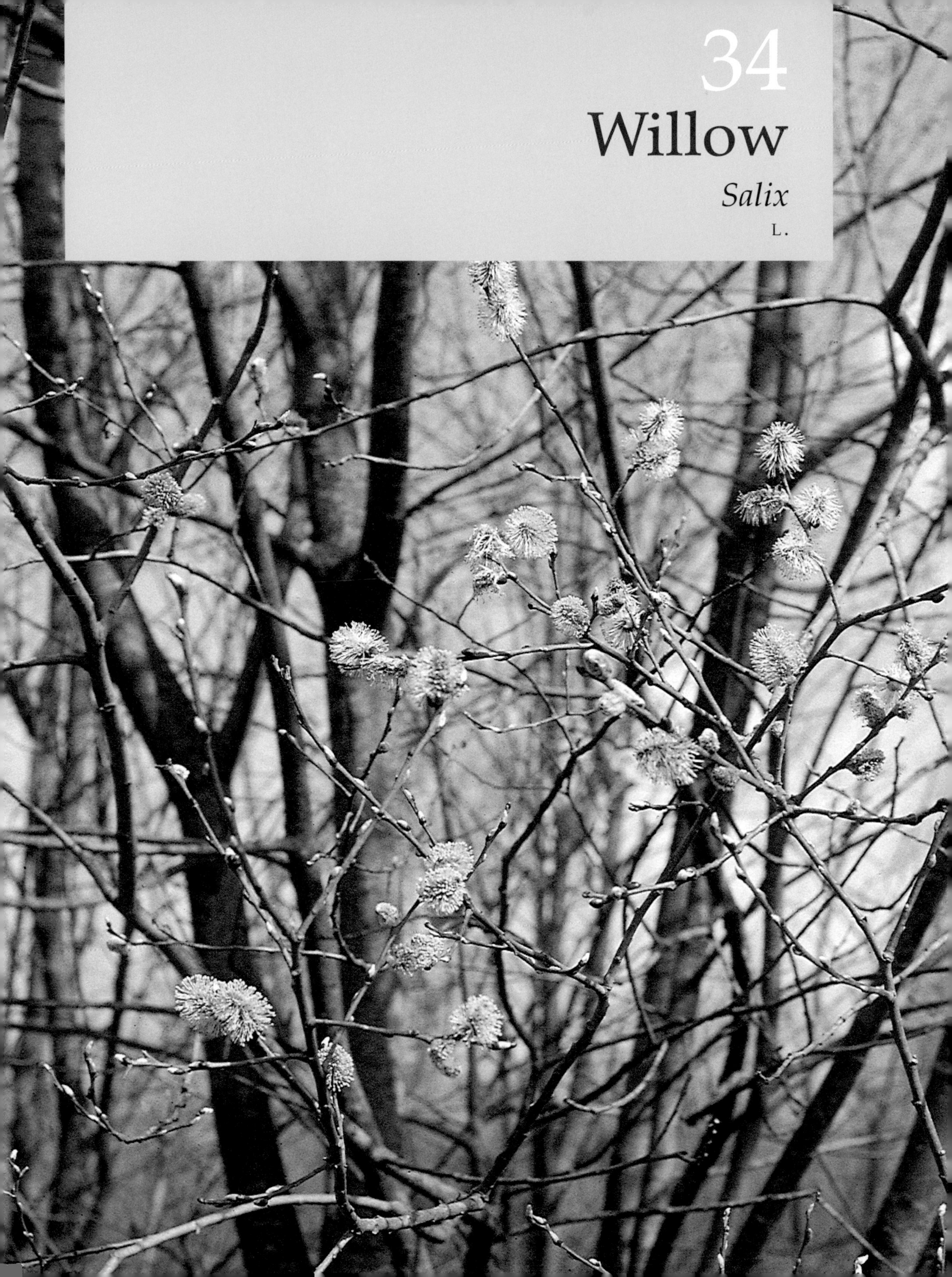

Willow

Salix L.

Salix is probably from the Celtic *sal* meaning 'near' and *lis* meaning 'water'. The species names are as follows: *alba* means 'white' in Latin; *caprea* is from *capra*, a she-goat (although *caprea*, the species name for goat willow actually means in Latin a female roe deer); *fragilis* means 'brittle'; sallow is from *pallidus*, the Latin word for 'greenish and pale'. 'Willow' itself is from old English, *welig*.

Salix × *rubens*

Salix cinerea subsp. *oleifolia*

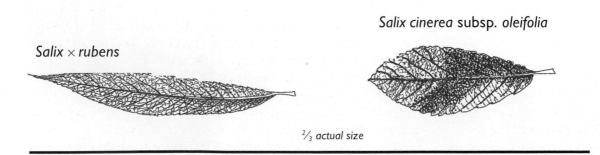

⅔ *actual size*

▲
Goat willow can produce a substantial trunk but the timber has little value.

Origin and distribution

There are three distinct groups of larger willows: tree willows, sallows, and osiers. Strictly speaking, the osier tribe is limited to a small number of plants, but commercial osier growers also use selections from the tree willows and the sallows. The term osier is used here in the commercial sense, meaning for the manufacture of baskets. Tree willows are seldom seen in plantations; in fact, they will not tolerate shade of any sort, and cannot be grown close together. The cricket bat willow, *Salix alba caerulea*, is a variety of white willow and is always planted at wide spacing on only the most fertile damp sites. The other tree willows are the almond-leaved *S. triandra*, the bay-leaved *S. pentandra*, and the crack willow, *S. fragilis*. The largest stem diameter of any willow in Britain in 1989 was a white willow, *S. alba*, at 235 centimetres (24 feet 2 inches in girth). The tallest was a cricket bat willow which reached 32 metres (105 feet) in 1985. The sallows are a large diverse group which includes the only real woodland willows; they can tolerate drier ground and some shade for a time, and reproduce freely by seed. The best known of these are goat willow, *S. caprea*, and common sallow, *S. cinerea* subsp. *oleifolia*. These are small trees with broad leaves when compared with the tree willow group. The largest common sallow was measured in 1991, and the ultimate size for the subspecies is 11 metres (36 feet) tall and 41 centimetres diameter (4 feet 3 inches in girth) at Wolford in Devon. There are much larger measurements of goat willow on record but these may be hybrids; the species is prone to hybridize and shows hybrid vigour. Pure sallow woods are rare, and usually confined to wet areas or places that have been devastated or suffered ground disturbance in some way. Willow seeds quickly blow in on the wind and produce a covering of trees seedlings. Sallows do not survive in woodlands for long, but they are often a permanent part of the roadside and wood-edge plant community. Until the nineteenth century withy beds for basket manufacture were common on damp sites everywhere especially in fenland districts such as the Somerset Levels. Several species and numerous clones of willow were grown to meet the various needs of the basket makers. Beds were planted at close spacing (55 000 per hectare) and harvested every year. The 'stools' would produce rods in great quantity for 10 years or so

▲
Common osier, like all willows, produces copious quantities of seed fluff in early summer.

▲
The goat willow has broad leaves and grows from seed on drier ground.

and then begin to decline as fertility ran out or rot set in. At this time the whole lot was usually grubbed out or grazed off with cattle and started again elsewhere using new cuttings.

Key features

Willows are a diverse genus, containing large trees, small trees, shrubs, and dwarf creeping plants. The most familiar are the silvery, narrow-leaved, waterside trees and round-headed pollards, or the rich green-leaved, urban, weeping willows. A consistent feature of all willows is their reluctance to maintain a single permanent leading shoot. Every spring a strong side bud will take over as the leading shoot, causing the tree to fork irregularly. In addition, vigorous plants will produce late sideshoots, which then grow on for such a long time that the growing tips die from the onset of winter cold. There are numerous combinations of shoot and foliage colour and leaf shape. A diagnostic feature of willow foliage, however, can be found in the winter buds, as each one has a single scale covering the whole bud. Trees are usually entirely male or female. The flowers are sessile catkins which, on many species, appear early in the year; on some (especially goat willow), male trees are quite spectacular. The flowers require pollination by insects and, because they are nectar-rich, are also good for bees. Seed is produced only a few weeks after the flowers, and it is minute and has only short viability. It is carried away by light fluff on the wind. Landing on a suitable cleared damp seedbed is a matter of some urgency, and rain soon afterwards is vital to press down the seeds into the soil.

◄

Cricket bat willow leaves have a bluish tint.

▲

Hybrid tree willows (*Salix × rubens*) at Huntingdon.

Growth conditions

At the present time none of the willows is planted very much as woodland, but they occur naturally as they have done since the end of the last Ice Age. In the British Isles there are willow species perfectly suited to every kind of natural (and artificial) environment. Willows are not particularly wind-firm, but falling over and regrowing are part of the natural life cycle of many species, particularly the crack willow, whose branches audibly crack off in a strong wind. Willows collectively provide shelter and food for a huge number of organisms, probably more than any other native tree species in Britain. Willows suffer defoliation due to rusts of the *Melampsora* species. Anthrancnose caused by the fungus *Marssonina salicicola* is a common disease on some weeping willows.

Propagation and management

Propagation by seed is difficult because of the short viability period. In addition not all species are fertile, and most trees have the sexes on different trees and are consequently useless in isolation. Many species cross-breed, and unwanted hybrids of dubious origin may be a problem for the grower and the plantsman. Propagation from seed has been achieved by cutting down fruiting twigs, pinning them down to damp bare ground, and providing irrigation. Winter cuttings are a more usual method of propagation, but some species, notably the goat willow, do not respond well. Rooted plants can be raised from summer cuttings under glass with mist and heat, but they are expensive to produce. Planting out is not without difficulty either, as bare-rooted sets often have extensive root systems that are inevitably damaged in the process. Containerized plants often become pot-bound or grow out of the bottom of the pot. Cuttings directly planted out have to be protected from frost heave, drying out, weed competition, and browsing animals. Potential sites require very careful selection because willow roots are widespreading and seek out underground water, particularly drains, which may soon be blocked. Seed fluff from female clones can also be a source of annoyance. It must be remembered that, once established, some willows are not easy to remove.

Timber quality and uses

Willow timber is resilient, light, stable, soft, and difficult to fracture. It takes impact very well but dents easily. It is white or pale buff in colour with increasing warm brown tints towards the centre of the tree. Sallow wood, although hardly ever used, may have streaks of red or yellowish-brown, and is very attractive. Sallow cambium is sometimes pink, but unfortunately the timber never picks up this bright colour. Willow wood is not durable and requires treatment if it is to be used in contact with moisture. Traditional uses include artificial limbs, slack cooperage and veneers for baskets, punnets, and similar items. Cricket bats are made from one selected type of willow, but the timber must be faultless. Watermark disease, caused by the bacterium *Erwinia salicis*, spoils many cricket bat willow plantations. Pollards were also retained and cut for rough posts and firewood. The advantage of pollarding was that livestock could be run on the ground below the trees without causing any damage. A new 'instant' pollard was simply a trimmed branch from an old one, pushed into the ground. The huge quantity of biomass produced by short-rotation willow coppice (12 dry tonnes per hectare per year) is being studied as a possible renewable source of fuel. Baskets made from osier rods have always been an important outlet for willow produce.

▲
White willow timber can be coloured from pure white to warm buff.

Yew

Taxus baccata L.

The scientific name *Taxus* is the old Latin name for the yew tree, perhaps from the Greek *taxon*, or 'bow'. *Baccata* means 'berried'. The common name is from the Welsh *yw*.

²⁄₃ *actual size*

Origin and distribution

Native to Europe, Iran, and Algeria, the yew is also widely distributed in Britain, particularly on lime-rich and chalk soils. It occurs in scattered populations in woodland especially under oak. Occasional small pure stands of yew can be found, which have probably arisen from the colonization of neglected pasture. The largest and most commonly encountered specimens are without doubt in churchyards. Many yews are enormous, but so broken up and hollow that they are difficult to measure. The largest stem diameter on record is at Defynnoc in Wales at 342 centimetres (35 feet 3 inches in girth) at ground level. The tallest tree on record in 1987 was at Belvoir Castle, where the quite exceptional height of 29 metres (95 feet) had been reached.

Key features

Yew is poisonous to humans and domestic animals; wilted foliage and seeds are particularly toxic. With their dense, sombre, evergreen rounded heads of branches, the yews are the oldest intact trees in Britain. Some yews are claimed to be over 3000 years old, but this is impossible to prove because old specimens are always rotten in the centre. They may be hollow or filled with damp

▲
Huge trees occur but the stems tend to become fluted.

black earth in which other plants, notably elder, thrive. An old yew in a Northamptonshire churchyard has put on only 3 centimetres (1 inch) of diameter in 45 years. Such slow growth is not a feature of young vigorous trees, however, and 4-millimetre-wide annual rings (5 rings to the inch) may be sustained on average through the first 60 to 80 years. Old trees are untidy, usually with a proportion of dead, curving, slender, branches, which are retained for many years and become very hard. The bark on the stem is deep red-brown to purplish-grey, flaking off in thin plates often to reveal bright crimson patches beneath. The foliage consists of spirally arranged solitary needles, spreading in two horizontal ranks on side shoots. The needles are very dark green, lustrous or matt above, and paler below with two bands of dull green stomata. The foliage colour and habit varies from one seedling to another. The shoots are green for several years due to curving leaf bases. Flowers occur in March, and there are separate male

▶
Yew berries are unique and easily
recognized.

and female trees. Numerous male flowers are clustered below the previous season's shoots, and they are pale yellow and shed clouds of pollen on warm dry days. Female flowers are tiny and are clustered near the tips of short shoots. Fruits are ovoid about, 12 millimetres ($\frac{1}{2}$ inch) across and consist of a green-to-olive black seed in a green cup, which is surrounded by a translucent bright-red albumen, which is open at the apex.

Growth conditions

The yew prefers lime-rich soils and is very much at home on chalk. It is shade-tolerant and will even survive under the heavy shade of beech trees. It is totally hardy but grows best in sheltered places. It is wind-firm and only rarely blows down; fallen trees are usually cut back and allowed to grow again. A huge specimen in Selborne churchyard that fell in the 1987 storm was actually winched back up into its original position and lived for a few years more. In exposed places, such as the Lake District fells, trees are stunted, rugged, and windswept. There are signs that yew is becoming increasingly intolerant of air pollution, and trees downwind of major industrial areas frequently have thin foliage due to reduced needle retention.

Propagation and management

There is almost no commercial management of yew in Britain at the present time, but the tree has potential as a high-quality timber producer. Propagation by seed can be accomplished by sowing it in open ground in March. Seed must be gathered 16 months before sowing and stratified in sand or stored in a cold store. Young seedlings must be protected from direct sunlight, and plants must be either lined-out or undercut each year that they remain in the nursery to encourage a fibrous root system. An alternative to seed is vegetative propagation using cuttings of semi-ripe wood (terminal shoots) taken in July, to be raised under glass with bottom heat and mist. Rooted cuttings should be fit for transplanting by the following October. Plants for hedging should always be grown from cuttings selected from bushy trees showing good hedge potential.

Timber quality and uses

Yew timber is hard and heavy, often with twisted grain and a tendency to split. It must be worked with great care. It is a rich chestnut brown with streaks of orange and even purple, and the narrow sapwood is pale cream. Epicormic shoots on the stem give rise to clusters of very decorative pin knots in a matrix of fine-textured, smooth, lustrous wood. Wide boards are infrequent because the tree has a habit of fluting and growing unevenly, often producing deep vertical recesses at the base of each branch. Yew wood finishes well and takes a fine polish. It has been used for all kinds of ornamental work, veneers, and small furniture. Sculptors use it to good effect, often highlighting the contrast between the heartwood and sapwood. Stained black, it is known as 'German ebony'. The yew is grown for high-quality hedges, and has a long history of topiary work. The oldest trees planted in churchyards must predate the Christian use of the sites, and probably have ancient religious significance associated with being evergreen. Being poisonous, yews growing in churchyards ensure that domestic animals do not graze there. The particularly dark and sombre Irish fastigiate yews have been popular since the cultivar was discovered in County Fermanagh in 1780. There are numerous other ornamental forms, including some with golden foliage and one with golden fruits.

▲
The colour and hardness of yew wood is outstanding, but it is by no means easy to work.

Glossary and Further Reading

acuminate (leaf) Finely pointed at the tip.

amorphous peat Structureless peat showing no signs of its plant origin.

anaerobic Oxygen-free conditions.

annual rings Rings of annual growth in stems and branches.

anthers Pollen-bearing parts of flowers.

auricled Having ear-shaped appendages.

axil The junction between the leaf stalk and the stem.

brackish Salty, as in river water near the sea.

bract A modified, reduced leaf which subtends a flower or inflorescence.

bractiole A small leaf-like attachment to a flower stalk between a true bract and the base of the flower.

broadleaf A broadleaved (as opposed to needle-bearing) tree.

cambium A layer of growing cells between bark and wood.

carpel The structure that encloses the ovules in a flowering plant.

calyx The outer part of the flower, usually green and protecting the bud.

clearfelling cycle The normal period of time taken for a plantation to pass peak productivity and be harvested.

conifer A cone-bearing tree, also includes yew here.

coppice Re-growth from the cut stumps of certain trees. Usually refers to a system of woodland management that involves repeated cutting on a short rotation.

crown (of a tree) The head of branches and all the foliage.

cuneate (leaf) Wedge-shaped with the point towards the base.

cupules A cup-like sheath surrounding some fruits.

dbh Diameter at breast height. The point from which a tree trunk is commonly measured, 1.3 metres (4 feet 3 inches) above ground level.

deciduous Keeping foliage for only one season.

deltoid (leaf) Triangular (also applies to cone, scales, and bark).

dioecious Plants in which the female and male reproductive organs are on different individuals.

edaphic The physical, chemical, and biological qualities of the soil that directly affect the growth of trees.

epicormic Shoots that sprout from the main stems of some trees a considerable time after normal branching has taken place.

epiphyte Plant with no roots that lives on rain water and organic debris, supported by tree trunks etc.

evergreen Holding needles or leaves for more than one year.

felted Having dense hairs like felt.

flushing Appearance of new foliage at the start of the growing season.

foliage	Leaves, buds, twigs, and shoots.	ovate	Oblong to elliptical, broadest at one end, resembling the longitudinal section of an egg.
glabrous	Without hairs.		
glaucous	Silvery-grey or blue-grey, often as a bloom on green foliage.	ovoid	Egg-shaped.
		ovule	Female reproductive structure containing the egg, which may participate in sexual fusion.
gley	Soil that is within the zone of a fluctuating water table showing a grey and brown mottled effect, caused by ferric iron being converted to grey ferrous iron in the absence of air.		
		pectinate	Shaped like a comb.
		petiole	The stalk of a leaf.
		pinnate	A compound leaf with leaflets on opposite sides of a central midrib.
globose	Rounded or roughly spherical in shape.	pollard	A tree that has its branches cut off at regular periods to produce small round-wood. The cuts are made above the level at which cattle can browse.
inflorescence	A head of flowers.		
internodal branching	On conifers, the random branches produced between each node or annual whorl of branches.		
lammas growth	A flush of summer growth.	provenance	The geographical region from which seed is collected.
lanceolate (leaf)	Narrow.		
lenticels	Breathing pores on shoots and twigs.	pubescense	Short soft hairs.
		pulvini (pulvinus)	The sharp pegs to which spruce needles are attached on the twigs.
mast	Heavy seed crops used for animal feeding, such as those of oak and beech.		
		raceme	An inflorescence (head of flowers) consisting of a main axis with several side branches of about equal length.
mast year	A year during which large quantities of seed are produced.		
medullary ray	Radial extension of the pith penetrating the primary tissues of the stem.	rachis	The central midrib of a pinnate leaf.
medullary sheen	Light reflected from medullary rays and the radial face of cut timber.	radial sheen	The shine on timber when it is cut or split through the centre of the tree.
monopodial	Having a single persistent trunk.	ring shake	A timber defect where annual rings split apart under stress.
mucronate	Abruptly terminated by a short stiff point.	sepals	The whorl of floral organs (usually green) outside (behind) the petals.
naturalized	A tree that has been introduced but has become well enough established to regenerate naturally.		
		silvicultural	Concerned with the cultivation of trees for timber or wood production.
needle	A narrow leaf on a conifer, often stiff and sharply pointed.	silviculture	The cultivation of trees, usually in woods and forests, primarily for the production of timber.
obtuse	Having a blunt or rounded apex.		

slack cooperage	Barrels that are not watertight, for powders or solid items.	strobile (strobilus)	A small cone-like structure.
spiral grain	Fibres spirally aligned in the tree with regard to the axis.	suborbicular	Half round, or less than a complete orb.
stag-headed	A tree with bare dead branches protruding from the top of the crown.	symbiotic	A mutually beneficial association such as that of fungal mycelium (the threadlike body of a fungus) with the roots of higher plants.
star shake	A timber defect where stress splits occur radially from the centre of a tree, usually forming a star.	tiered branches	Flat branches held in layers one above the other.
stipule	A leafy bract at the base of an ordinary leaf stalk; stipules usually occur in pairs.	truncate	A leaf that is squared off at the apex.
stomata	Breathing pores in a leaf.	undercut	A nursery technique where roots of plants in beds are shortened by drawing a blade at a set depth below them.
stored coppice	Coppice that is retained to produce trees again. This may be singled out to improve quality and speed of growth.	underplanting	Establishing a crop of shade-tolerant trees under the light cover of existing trees.
strainers (fencing)	Large posts at the corners or mid-way along fencelines from which wires are strained.	underwood	Shrubs growing under the shade of large forest trees.
stratification	A method of propagation to break seed dormancy, for example pre-chilling before sowing.	variety status	A botanical rank subordinate to species, a morphological variant.

Further reading

Bevan, D. (1987). *Forest insects (FC Handbook No. 1)*. HMSO, London.

Corkhill, T. (1984). *A glossary of wood*. Stobart & Son Ltd, London.

Evans, J. (1984). *Silviculture of broadleaved woodland* (FC Bulletin No. 62). HMSO, London.

Forestry Commission (1993). *Forestry facts and figures 1992–93*. HMSO, London.

Gruffydd, J. St. B. (1987). *Tree form, size, and colour. A guide to selection, planting, and design*. E. and F.N. Spon Ltd, London.

Harrison, C.R. (1975). *Ornamental conifers*. David & Charles, Newton Abbot, Devon.

Heywood, V.H. (ed.) (1978). *Flowering plants of the world*. Oxford University Press, Oxford.

Hillier Nurseries (Winchester) Ltd (1993). *Hillier's manual of trees and shrubs*. Sixth edition. David & Charles, Newton Abbot, Devon.

James, N.D.G. (1955). *The forester's companion*. Basil Blackwell, Oxford.

Johnson, H. (1979). *The international book of wood* Mitchell Beazley Ltd, London.

Lines, R. (1987) *Choice of seed origins for the main forest species in Britain* (FC Bulletin No. 66). HMSO, London.

Mitchell, A.F. (1974). *A field guide to the trees of Britain and northern Europe*. William Collins Sons & Co, Glasgow.

Mitchell, A.F., Hallett, V., and White, J.E.J. (1990). *Champion trees in the British Isles*. HMSO, London.

Mondadori, A. (1982). *The Macdonald encyclopedia of trees*. Macdonald & Co Ltd, London.

Peterken, G.F. (1981). *Woodland conservation and management*. Chapman & Hall, London.

Phillips, R. (1978). *Trees in Britain, Europe, and North America*. Pan Books Ltd, London.

Rackham, O. (1990). *Trees and woodland in the British landscape*. J.M. Dent & Sons Ltd, London.

Ratcliffe, D.A. (1984). *Post-medieval and recent changes in British vegetation: the culmination of human influence*. New Phytol. **98**, 73–100.

Rushforth, K.D. (1987). *Conifers*. Christopher Helm Ltd, Bromley, Kent.

Sheat, W.G. (1948). *Propagation of trees, shrubs and conifers*. Macmillan & Co Ltd, London.

White, J.E.J. (1981). *Hazel coppice at Westonbirt*. Forestry Commission, Edinburgh.

Index

Page numbers in Roman numerals refer to the Introduction. Italic numbers denote reference to illustrations.

'Scots mahogany' 5
seed ferns xi
Selbourne 208
Sequoia sempervirens 175–80
Sequoiadendron giganteum 175–80
ship building, *see boat building*
silver birch, *see Betula pendula*
Sitka spruce *see Picea sitchensis*
   see also 11, 70, 76, 82, 183–4,
      186
small-leaved lime, *see Tilia cordata*
southern beech, *see Nothofagus*
Speech House Arboretum xv
spruce, *see Picea*
   Norway, *see Picea abies*
   red 183
   Serbian, *see Picea omorika*
   silver 192
   Sitka, *see Picea sitchensis; see also*
      11, 70, 76, 82, 183–4, 186
squirrels 23, *87*, 87, 131, 197
Stapehill, Dorset 63, 65
Stilton Fen, Cambs. 129
stock knife 6
storms xix, xx
sugi 180
Surrey heath *33*
swamp cypress xiv
sweet chestnut, *see Castanea s
   ativa*
sycamore, *see Acer pseudoplatanus*

taiga xii
tanning and tanbark 53, 131
Taxus baccata 205–9
Thetford Chase *41*
Thuja plicata 37–42
Thunberg, Carl · xv
Tilia cordata 115–20
   × *europaea* 117, 119
   platyphyllos 117
Tortrix viridana 142
Tradescant, John xiv
tree
   creeper 71
   definition of x
   of Heaven xv
   of life 38
   shelters 40
Triassic Period x
Tsuga 68
   canadensis 92
   heterophylla 91–6
tulip tree xiv

Veitch, John Gould xv, 111
   nursery 177
voles 100

weathered sycamore 198
Wellington, Duke of 176

Wellingtonia 176
western hemlock, *see Tsuga
   heterophylla*
western red cedar, *see Thuja plicata*
Westonbirt 9, 11, *47*, *75*, 135,
   178
white deal 186
whitewood 186
wild cherry, *see Prunus avium*
willow, *see Salix; see also* 129,
   149
   crack, *see Salix fragilis*
   cricket bat 204
   goat, *see Salix caprea*
Wilson, Ernest xv
Wistman's Wood, Dartmoor 129
woodland x
wood pasture 99, 101
Wychwood 123
Wykeham Moor 11

Xanthomonas populi 166

yew, *see Taxus baccata*
   golden 209
   Irish 209

zones, forest xii